The Allegheny Woodrat

John D. Peles · Janet Wright
Editors

The Allegheny Woodrat

Ecology, Conservation and Management of a Declining Species

Springer

Editors

John D. Peles
Biology Department
Pennsylvania State University-McKeesport
McKeesport, PA, USA
jdp12@psu.edu

Janet Wright
Biology Department
Dickinson College
Carlisle, PA, USA
wright@dicksinson.edu

ISBN 978-0-387-36050-8 ISBN 978-0-387-36051-5 (eBook)
DOI 10.1007/978-0-387-36051-5

Library of Congress Control Number: 2008926524

Printed on acid-free paper

springer.com

To Kim, Joshua, and Mikayla with love and thanks for your patience and support

JDP

To my many students who have trapped, tracked, analyzed, and admired the Allegheny woodrat

JW

In many places throughout these mountains great slides of broken boulders form bare patches on the slopes. Wherever these conditions are sufficiently extensive, one is likely to find signs of Neotoma.

It has strikingly large ears, bright and prominent eyes, and above all, conspicuous long whiskers, the seat of a highly developed tactile sense, which guides the animals in their quick and certain movements among the hazardous ledges and cliffs that they travel, often in total darkness. They are exceedingly agile while traveling over rocks. At Bake Oven Knob they habitually ascend and descend [a] crevice, leaping from one side to the other at top speed.

Passages to the nests are frequently barricaded with a miscellaneous assortment of sticks, stones, leaves, and other objects, doubtless designed to baffle the larger enemies. Frequently, at some distance from the nest, a heap of objects equally miscellaneous and of considerable bulk is stored. I believe that the more desirable storage food is kept in or near the nest, while the large heaps of refuse found at a distance from the nest may be reserve supplies of less desirable food material, or simply objects that have been carried there in fulfillment of the animal's evident desire to store objects in heaps.

E. L. Poole. 1940. A life history sketch of the Allegheny woodrat. Journal of Mammalogy 21:249–270.

Foreword

The Allegheny woodrat (*Neotoma magister*) once ranged from southern Connecticut westward to Indiana and southward to northern Alabama. Now extirpated or declining in over 35% of its range, the species appears to be relatively secure only in Kentucky and in parts of Virginia and West Virginia. Given that until relatively recently the Allegheny woodrat was considered to be a subspecies of the eastern woodrat (*N. floridana*), much of our general knowledge of woodrats in eastern North America is based on research conducted in localities now within the range of *N. floridana*. However, studies over the past 25 years have shown that the Allegheny woodrat differs from its close relative. For example, it is ecologically distinct, preferring caves and other habitats with substantial and structurally complex rock cover. Thus, a volume devoted to the Allegheny woodrat has been critically needed in order to assess our knowledge of this species and determine how this information can be used to develop appropriate strategies to ensure its survival.

Beginning in the 1980s, a group of volunteer mammalogists assisted the Pennsylvania Game Commission in monitoring Allegheny woodrat populations throughout the state. At my assigned site, Strangford Cave in Indiana County, I typically captured 4–6 animals. Yet, a former graduate student in my department, who had worked at Strangford in the 1970s, estimated a population size of 40–50. Unfortunately, the history of this single site has been repeated many times throughout much of the species' range. A decline in populations of *N. magister* was first noted in eastern Pennsylvania by John Hall in the 1980s. With Allegheny woodrats now having disappeared from eastern Pennsylvania and in active decline across the state, it is particularly appropriate that this volume was initiated by two Pennsylvania mammalogists.

The content of this book summarizes what is known about the ecology of the Allegheny woodrat, with an emphasis on reasons for its decline, as well as application of this information to its conservation and management. The initial two chapters provide a framework for the sections that follow. The first reviews information on history and current status throughout the range, while the second presents a synergistic model to explain the species' decline. The second section of the book includes a series of chapters on ecology of the Allegheny woodrat, from landscape and habitat-scale issues to life history characteristics at the local population level. In each case, whether it is population dynamics, microhabitat selection by foraging woodrats, den selection, or dietary preferences, the authors speculate on how these ecological characteristics might interplay with disturbance to either limit or cause the decline of populations. The third section of the book deals with conservation and management issues, starting with descriptions of long-term monitoring programs in various states. Discussions of genetic diversity patterns and their relevance to management, tailoring of management strategies to biological and ecological characteristics, and review of information bearing on the potential for reintroductions are included here. Another chapter considers how information from *N. magister* can be applied to conservation of other woodrat species. The final section outlines future research needs, as well as summarizing how knowledge gathered during more than two decades of research on Allegheny woodrats can be used to help conserve this species. Although it is hoped that the information compiled in this volume will contribute to actions that will slow or halt the decline of the Allegheny woodrat, conservation biologists may also find this species to be of interest as a general model for analysis of the decline process. Because *N. magister* populations exhibit a continuum from extirpation to apparent health, comparative ecological information gathered on these various populations may help to develop generalizations that can be applied to other species in decline.

Our native small mammals are a beautiful and fascinating component of our terrestrial biodiversity. Unfortunately, rats and mice suffer from a public relations problem, perceived as lacking the charisma, and therefore lacking the enthusiasts associated with some other taxa. A group of Pennsylvania woodrat researchers once made a tongue-in-check proposal that the common name of the Allegheny woodrat should be changed to "long-tailed woods bunny." Fortunately, more and more people are accepting the notion that maintaining biodiversity is a "good thing," even if it is difficult to make a case for the critical importance of any single species.

In asking ourselves what would happen if a particular species disappeared, surely the answer is that "it depends." The loss of the American chestnut sent a shock wave of ecological adjustments through our eastern forests. In fact, it has been speculated that this event set the stage for the decline of *N. magister*. Certainly, small mammals such as the Allegheny woodrat are important contributors to the prey base for larger mammals and raptors. But perhaps, the only thing we can say with certainty is that the loss of any species leaves us poorer, whether it is the loss of the passenger pigeon that once darkened our skies or of the most secretive species that share our earth. The biologists who contributed to this volume have not only spent significant amounts of time conducting research on Allegheny woodrats, but also have contributed untold volunteer hours to help in ensuring their survival. Writing in *A Sand County Almanac*, a work that has inspired several generations of conservationists, Aldo Leopold asserted that "There are some who can live without wild things and some who cannot." This book is clearly the product of those who cannot.

Indiana, PA Alicia V. Linzey

Preface

In the early 1980s, the Pennsylvania Biological Survey—a research advisory body to the Pennsylvania Game Commission—recommended two priorities regarding the Allegheny woodrat (*Neotoma magister*): (1) surveys to determine the present distribution of the species; and (2) investigations to understand its ecological requirements. These recommendations reflected the magnitude of the task facing biologists in Pennsylvania and many other states at that time. The task was to document the extent and progress of the decline for a species about which little was known. Implied in the recommendations also was the need to understand the reasons for the decline of the Allegheny woodrat.

For nearly two decades, biologists employed by state agencies devoted many hours to documenting the extent of the Allegheny woodrat's distribution and monitoring the status of existing populations. These were pioneering studies, in that the status of nongame species was a relatively new focus for these agencies. While biologists from academic settings assisted with the monitoring process, most of their efforts were focused on understanding the systematics, ecology, and natural history of woodrats—the business of classical mammalogy. Although a great deal of potentially valuable information was gained from the efforts of both groups, much of it remained unpublished or available only through reports to state agencies. In some cases, this was due to time constraints faced by biologists within those agencies. In other cases, the lack of published information on *N. magister* reflected the difficulties in collecting statistically rigorous data in small populations that are subject to fluctuations in time and space. In fact, the first publications reporting empirical research on woodrats did not appear until very recently as researchers in West Virginia, where populations are relatively stable, began to take interest in this species.

In September 2005, a workshop on the Allegheny woodrat—organized by Jerry Hassinger—was held at Dickinson College to facilitate communication among researchers and field personnel about the status of our knowledge of this species. Such conferences are useful, but one of us (JP) came away with a conviction that there was a large, diffuse body of information that begged to be summarized in book form. From this idea developed our partnership to coedit the project. As mammalogists who have devoted a great deal of time and energy to the study of the Allegheny woodrat, initially our primary purpose was to synthesize current knowledge of the species. Thus, this book answers questions such as: What is the current distribution and status of *N. magister*? What factors have contributed to its decline? What do we know about its ecology and genetics? What is the future of conservation and management for this species?

In this book, we use the term *decline* to refer to a long-term trend of decreasing abundance and distribution of a species within all or part of its historical range. As we considered the implications of this definition, we realized that this term likely applies to many species in North America and elsewhere. Thus, much of what we have learned from the study of the relatively obscure Allegheny woodrat has the potential to make a valuable contribution to the field of conservation biology. Consequently, the second purpose of this book is to highlight general principles that may be applied to the study of other declining species. In fact, we were encouraged by a reviewer of the original book prospectus to include a chapter on the status of other species of *Neotoma* based on the knowledge that several species or subspecies within this genus have experienced decline.

Most of the contributors to this book represent researchers currently involved in the study of the Allegheny woodrat. In some cases, authors who have previously published on a topic have integrated their own work with other published and unpublished data to provide a synthesis of our current knowledge on that topic. Other chapters represent the synthesis of data on a topic that has not previously appeared in the published literature. In fact, a particular goal was to use this book as an opportunity to present previously unpublished data that contribute to the synthesis of our current knowledge on the topic. All chapters were peer reviewed and underwent significant revision following the review and editorial processes. Our goal has been to bring together what has been a disparate, diffuse patchwork of information into a coherent, integrated picture. Our hope is to provide both good models and some caveats from hard lessons learned for those approaching these same questions for other species in decline.

We are grateful to Janet Slobodien and Ann Avouris at Springer for their support of this project and their assistance in the publication process. Several external reviewers of the prospectus for this book provided helpful suggestions. George Feldhamer, Mark Ford, Alicia Linzey, Kathleen LoGiudice, Carolyn Mahan, Joe Merritt, Gene Rhodes, Tim Smyser, and Petra Wood provided helpful comments on one or more chapters in the book. Most of all, we appreciate the spirit of the authors in their shared concern for the future of this unique animal.

Mckeesport, PA, USA John D. Peles
Carlisle, PA, USA Janet Wright

Contents

Section I History and Current Status

1 **History and Current Status of the Allegheny Woodrat** 3
Janet Wright

2 **Multiple Causes of the Allegheny Woodrat Decline: A Historical–Ecological Examination** 23
Kathleen LoGiudice

Section II Ecology

3 **Woodrat Population Dynamics and Movement Patterns** 45
Petra Bohall Wood

4 **Home Range, Movements, and Habitat Selection** 63
Steven B. Castleberry

5 **Den Use Behavior of Allegheny Woodrats Inhabiting Rock Outcrops in Pennsylvania** 75
John D. Peles and Janet Wright

6 **Food Selection and Caching Behavior** 93
Nikole L. Castleberry and Steven B. Castleberry

Section III Conservation and Management

7 **Lessons from Long-Term Monitoring of Woodrat Populations** 109
Michael T. Mengak, Calvin M. Butchkoski, Daniel J. Feller, and Scott A. Johnson

8 Managing Surface Rock Communities for *Neotoma magister* 133
Jerry D. Hassinger, Calvin M. Butchkoski, and Duane R. Diefenbach

9 Genetic Diversity of Woodrats: Implications for Conservation 153
Timothy J. Smyser and Olin E. Rhodes, Jr.

10 Reintroduction of Woodrats: Concepts and Applications 169
Thomas L. Serfass

11 Status and Conservation of Other At-Risk Species of North American Woodrats 193
George A. Feldhamer and Aaron K. Poole

Section IV Synthesis

12 Synthesis: Lessons Learned and Recommendations for the Future ... 209
John D. Peles and Janet Wright

Index 225

Contributors

Calvin M. Butchkoski
Pennyslvania Game Commission, Harrisburg, PA 17110

Nikole L. Castleberry
Georgia Department of Natural Resources, Nongame Conservation Section, Social Circle, GA 30025

Steven B. Castleberry
Warnell School of Forestry and Natural Resources, University of Georgia, Athens, GA 30602

Duane R. Diefenbach
U.S. Geological Survey, Pennsylvania Cooperative Fish and Wildlife Research Unit, Pennsylvania State University, University Park, PA 16802

George A. Feldhamer
Department of Zoology, Southern Illinois University, Carbondale, IL 62901

Daniel J. Feller
DNR-WHS Natural Heritage Program, Appalachian Laboratory, University of Maryland, 301 Braddock Road, Frostburg, MD 21532

Jerry D. Hassinger
Pennsylvania Biological Survey, 105 Adams Lane, Millersburg, PA 17061

Scott A. Johnson
Indiana Department of Natural Resources, 553 East Miller Drive, Bloomington, IN 47401

Alicia V. Linzey
Department of Biology, Indiana University of Pennsylvania, Indiana, PA 15705

Kathleen LoGiudice
Department of Biological Sciences, Union College, Schenectady, NY 12308

Michael T. Mengak
Warnell School of Forestry and Natural Resources, University of Georgia, Athens, GA 30602

John D. Peles
Pennsylvania State University, Greater Allegheny, McKeesport, PA 15132

Aaron K. Poole
Department of Zoology, Southern Illinois University, Carbondale, IL 62901

Olin E. Rhodes, Jr.
Department of Forestry and Natural Resources, Purdue University, West Lafayette, IN 47906

Thomas L. Serfass
Department of Biology, Frostburg State University, Frostburg, MD 21532

Timothy J. Smyser
Department of Forestry and Natural Resources, Purdue University, West Lafayette, IN 47906

Petra Bohall Wood
U.S. Geological Survey, West Virginia Cooperative Fish and Wildlife Research Unit, Division of Forestry and Natural Resources, West Virginia University, P.O. Box 6125, Morgantown, WV 26506

Janet Wright
Department of Biology, Dickinson College, Carlisle, PA 17013

Section I
History and Current Status

The decline of few species has been as mysterious as that of the Allegheny woodrat (*Neotoma magister*). The chapters in this section show why understanding the disappearance of this species has been such a knotty problem. Reference to a species as "declining" implies documentation of a decrease in abundance and distribution within at least some part of the species' range. In the case of the Allegheny woodrat, this seemingly simple task was confounded by its uncertain taxonomic status. In Chapter 1, Janet Wright reviews the complicated taxonomic history of *N. magister* and discusses how this and other factors contributed to the delay in recognizing the decline of this species. She then provides a review of the current distribution of this species and its conservation status as designated in the states of its range.

Once the decline of a species is recognized, the next step is to understand the underlying causes. For many years, biologists used a single-factor approach to explain the decline of the Allegheny woodrat. Kathleen LoGiudice (Chapter 2) provides a review of the main hypotheses to explain the decline, and reviews the shortcomings of each of these single-factor hypotheses. She presents a convincing case for considering interactions among multiple factors within a historical framework to explain the woodrat's decline. A careful examination of our attempts to understand the reasons for the disappearance of the Allegheny woodrat should provide a lesson to conservation biologists that context, including taxonomy and history, must be a part of the analysis of any species in decline.

John D. Peles
Janet Wright

Chapter 1
History and Current Status of the Allegheny Woodrat

Janet Wright

Introduction

In early 2007, the announcement of a new species of clouded leopard (*Neofelis diardi*) received international press coverage. In fact, this animal, the largest predator in Borneo, was already well known to biologists. The actual news was that DNA analysis showed it to be a species distinct from its close mainland Asian relative, a result that researchers said should "increase the urgency of clouded leopard conservation efforts" (Buckley-Beason et al. 2006). At the same time, another research group was reporting that the population size of clouded leopards in Borneo was considerably lower than previously estimated (Wilting et al. 2006).

The case of the clouded leopard illustrates twin problems that bedevil our understanding of species of conservation concern—identifying what the species is and recognizing its level of distress. However, most threatened animals have considerably less public visibility and research support than the clouded leopard. The Allegheny woodrat (*Neotoma magister*) may be a more realistic model for illustrating the significance of these problems, and how we can confront them for better conservation.

In this chapter, I first trace the convoluted taxonomic history of the Allegheny woodrat. While the case has some unusual twists and turns, similar confusion has occurred for many other species. I then review how the Allegheny woodrat came to be perceived as a species in decline, and why this perception has been hard to validate with data. Where baseline data are lacking, as is often the case, documenting a decline may require creative reconstruction of history by a variety of means. Our experience with the Allegheny woodrat is an object lesson in what basic information needs to be gathered before proceeding to management plans. This information includes an understanding of genetics and geography, attention to the primary literature (however old), use of the best baseline data available, establishment of clear

---

J. Wright
Department of Biology, Dickinson College, Carlisle, PA 17013

J.D. Peles, J. Wright (eds.), *The Allegheny Woodrat*, DOI: 10.1007/978-0-387-36051-5_1,

goals and criteria for meeting those goals, and development of a framework for coordination and communication.

As background for this review, and to aid in understanding the basis for the literature related to woodrat taxonomy, I compiled a database of existing museum specimen records for what is now designated *N. magister*. These records, from standard museums (Hafner et al. 1997), but also from additional sources cited in the literature or suggested by colleagues, typically included specimen identification number, collection date and locality, and often the collector's name. To select records for the database, I did not examine most of the specimens, but used recorded collecting locality as the primary guide. I first included all *Neotoma* specimens whose collection county appeared to fall within the geographic range mapped in Hall (1981) for *Neotoma floridana magister* (as it was then designated). I then examined records along the ambiguous southeastern border of that mapped range (through North and South Carolina, Georgia, and Alabama) and followed the determinations by Ray (2000) for geographic range of *N. magister* versus *N. floridana*.

The total museum specimen database (available on request) comprised over 950 records. From these, I produced maps of collecting localities, using geographic coordinates listed with the record or that I estimated according to Mammal Networked Information System (2007) guidelines. Specimens whose collecting locality was described only to the county level were excluded from the geographic data, except for two records that were the exclusive ones for that county (Chester County, PA and Buckingham County, VA); for these I estimated coordinates near the center of the county. I created time-series maps of the specimen localities using ARC GIS 9.2 (Environmental Systems Research Institute, Inc.) to illustrate the extent of documented geographic range of this species at different points in time.

The records in the museum specimen database, with their dates of collection, imply how much was known to biologists about *N. magister's* morphology and distribution at various points in the past. In comparing these records with the historic literature, I have made the simplifying assumption that the collection date of each specimen is when it became "known to science" and available to researchers. While this may not be true of all specimens, it is probably true of the considerable majority.

Early Recognition: Extant or Extinct?

The Allegheny woodrat was first described as *Neotoma magister*, with a later name *Neotoma pennsylvanica* being discarded as invalid. It was then reclassified as a subspecies of *Neotoma floridana (N. floridana magister)*, but more recently has been recognized as a distinct entity under the original name, *Neotoma magister* (Castleberry et al. 2006). Such a bald account, however, falls short of conveying how this confusing history evolved as a result of a gradually accumulating base of specimen material, beginning with a few fossil jaws.

Initially, the species was introduced to science by biologist Spencer Fullerton Baird. As Assistant Secretary to the Smithsonian Institution, Baird contracted biological teams to accompany the railroad survey crews in the American west, shipping specimens back to Washington, and forming the core of what is now the United States National Museum of Natural History. In 1857, on the basis of these specimens, Baird published a synopsis of known and newly described American mammals, including seven living species in the woodrat genus *Neotoma*, five of them new to science (Baird 1857).

Baird's geographically extensive North American collections allowed him to put into context some curious cave fossils he had brought with him to the Smithsonian from his years as a student and professor at Dickinson College in Carlisle, Pennsylvania (Miller 1940). Although they were incomplete jaw fragments, he recognized them as *Neotoma*. The only other eastern U.S. woodrat specimens Baird had for direct comparison were six skulls of *N. floridana* from South Carolina, Georgia, and Arkansas. Baird asserted that the Carlisle specimens were much bigger, and speculated that the animal they represented "could not have been less than 12 inches in length." He named the large species '*Neotoma magister*' and declared it extinct (Baird 1857).

Baird also commented on a woodrat collected in New York, along the Hudson River, which he had as a taxidermy specimen without a skull. Despite being impressed with its "unusually large size," Baird (1857) grouped it with the smaller *N. floridana* from the Gulf Coast, "although I have not heard of any intermediate localities." Why Baird chose to make this convoluted determination instead of the more parsimonious one of assigning the New York specimen to *Neotoma magister* is unknown.

Knowledge of the geographic range of North American woodrats accumulated slowly. Ironically, the next discovery in the northeastern U.S. was only 30 km from where Baird had done his early fieldwork. In 1893, two woodrats were collected on a rocky ridge south of Carlisle, Pennsylvania, and sent to the Academy of Natural Sciences at Philadelphia, where Witmer Stone, primarily known as an ornithologist, described them as a new species, *Neotoma pennsylvanica* (Stone 1893). Stone gave no indication he was aware of Baird's fossil *N. magister*. He differentiated his Pennsylvania woodrats from two Florida specimens of *N. floridana*, citing *pennsylvanica's* hairier tail, larger size, and several cranial characters (Stone 1893).

The following year, Samuel Rhoads (1894) attempted to resolve whether Baird's fossils and Stone's *N. pennsylvanica* were really different species, and whether they were distinct from *N. floridana*. Again, the investigation was constrained by limited specimen material. To represent *N. pennsylvanica*, Rhoads used Stone's two skins and skulls. For comparison, he did not have Baird's *N. magister* type specimens, but substituted "fossil" bones from two eastern Pennsylvania caves, which he assumed were equivalent to Baird's *N. magister*. Rhoads also used two woodrats that had been trapped in a cave in western Virginia, far from the known range of any woodrat to that date. Surprisingly, he included these as *N. magister* without comment, despite Baird's claim that *magister* was an extinct species and despite the fact that their collector, Philadelphia biologist E. D. Cope, had called them "*Neotoma floridanum*" [sic], a logical choice given the Mississippi drainage from which they were taken (Cope 1869). From this small sample Rhoads (1894) concluded that the

"*N. pennsylvanica*" specimens were indistinguishable from *N. magister*, and Baird's original name *magister* should take precedence for both living and fossil forms.

The "unusually large" Hudson River woodrat that had been grouped by Baird with southern *N. floridana* also got Rhoads's attention. It would have been logical to include this specimen with his other living *N. magister*, but Rhoads did not do so. He noted that all his northern (i.e., *N. magister*) specimens had been collected in caves or cave-like fissures, whereas the Gulf Coast *N. floridana* were not cave animals. In fact, he consistently referred to the northern species as "cave rat" and the southern as "woodrat." Apparently, this habitat distinction convinced him that Baird's Hudson River, New York, specimen had to be a "large *Neotoma floridana* ... probably imported in a cargo of southern lumber" (Rhoads 1894).

Rhoads's speculation about the anomalous Hudson River woodrat came under immediate challenge when Allen (1894) reported a specimen caught live along the Hudson, a few miles upstream from Piermont, New York. Allen described features showing his specimen was not *N. floridana* and claimed that his animal, as well as the Hudson River one seen by Baird, was "doubtless" Stone's *N. pennsylvanica*. The woodrat was caught along a "cliff, full of deep crevices" (Allen 1894), an observation consistent with Rhoads's claim that the northern species was a cave and fissure specialist.

In the same year, C. H. Merriam (1894) proposed a systematic arrangement of species of the genus *Neotoma*. He did not say what northeastern woodrat specimens he examined, but he was probably aware of those held in the Academy of Natural Sciences of Philadelphia and the U.S. National Museum. This would have included 17 specimens (from two localities in Pennsylvania and one each in Virginia, Kentucky, and New York) plus "fossil" bone deposits from several Pennsylvania caves (Fig. 1.1).

Merriam, like Rhoads, was a strong subscriber to the theory that each species was associated with a particular habitat. He completed a detailed study in Arizona (Merriam 1890) that led to the development of his enormously influential Life Zones concept, in which flora and fauna were arranged in discrete elevational bands from base to summit of western mountains. In the *Neotoma* review, he applied his Life Zone approach to the eastern woodrats, separating them into *N. pennsylvanica*, with a geographic range he estimated (from no more than five localities) as "Allegheny Mountain region of Pennsylvania and probably the whole of the southern Alleghenies; north to southern New York" and *N. floridana*, belonging to the "austroriparian fauna of South Atlantic and Gulf Coasts and upper Mississippi Valley" (Merriam 1894). Merriam's range designations thus emphasized a montane habitat for the northern species and a valley habitat for the southern one. As to what he termed the "subfossil" *N. magister*, Merriam noted that Baird had collected it in a Pennsylvania valley, which would make it difficult to reconcile with his supposedly montane *N. pennsylvanica*. He elected to keep *magister* as a separate (extinct) species, speculating that the cave *N. magister* of Pennsylvania were the same as Pleistocene bones from Missouri caves (Merriam 1894).

As new material accumulated, E. A. Goldman (1910) undertook a new revision of the entire genus *Neotoma*. To address the *magister/pennsylvanica* split, he

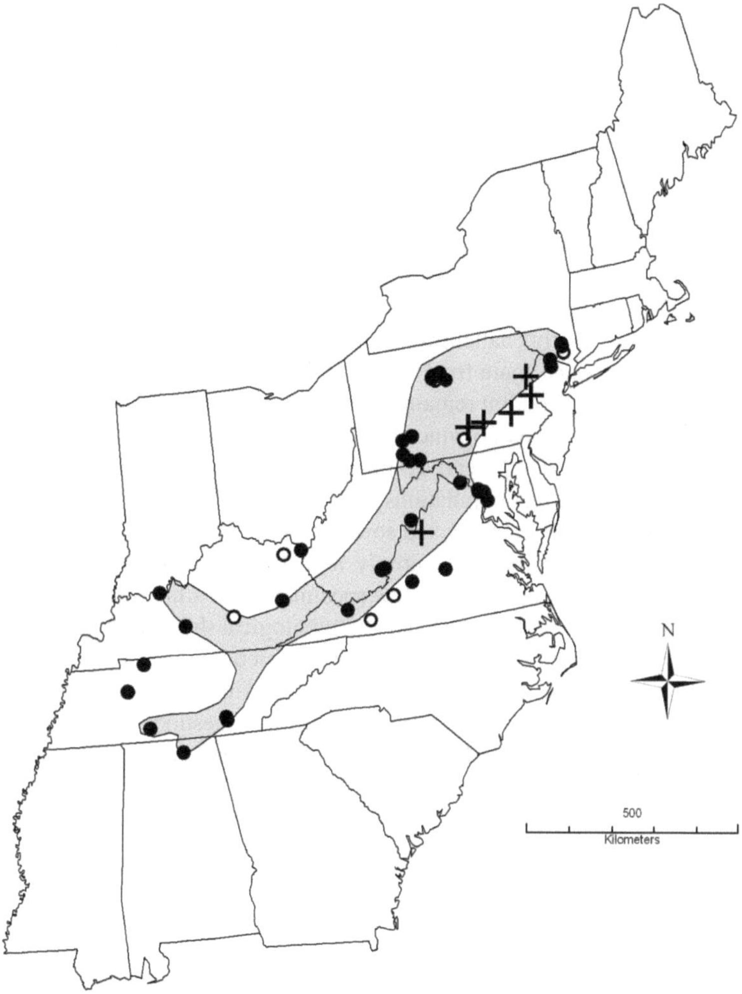

Fig. 1.1 Early collecting localities for museum specimens of *Neotoma magister*, with range of the species (shaded) as depicted by Goldman (1910). Open circles represent specimens collected by the time of Merriam's (1894) taxonomic revision. Closed circles represent additional specimens collected by the time of Goldman's (1910) revision. Crosses represent fossil specimens collected before 1910

took pains to obtain Baird's original type specimens. He devised skull metrics to compare them with *N. pennsylvanica*, using five new specimens from West Virginia, Kentucky, and the Potomac River near Washington. In a comparative table, the Baird mandibles were generally larger in all metrics than the "*pennsylvanica*" jaws, and Goldman concluded that *N. pennsylvanica* and *N. magister* should be separately recognized, but only after taking the questionable step of omitting two of Baird's mandibles because they were "probably *N. pennsylvanica*."

Goldman (1910) also published the first range map for "*N. pennsylvanica*." It was based on much more material than had previously been available—94 specimens from 18 localities. The map depicted a range extending from northern Alabama up the Tennessee River Valley in a narrow band northward along the Appalachians to the Hudson River in New York, and included a branch across Kentucky but excluded Ohio and Indiana (Fig. 1.1). Goldman did not comment specifically on this animal's relationship to *N. floridana*; but as he placed the two species far apart in his text and said that *N. pennsylvanica* required "no close comparison with any known living form," apparently he regarded it as easily distinguishable.

Goldman's scheme, with *N. pennsylvanica* and fossil *N. magister* as separate species, was not seriously challenged for three decades, but Poole (1940) eventually raised the question, once again from an expanding specimen base. Poole was particularly impressed with woodrat remains he had found alongside pre-Colonial human artifacts. His investigations convinced him that the woodrats present in Pennsylvania and Maryland, from the earliest postglacial deposits through the archaeological material to current-day populations, were all one continuous species. Poole reexamined Baird's type specimens and seven museum skulls of "*N. pennsylvanica*," concluding that they overlapped entirely and that the name *pennsylvanica* was invalid. Although Poole used only a few specimens in his measurements, his arguments carried the day, perhaps because the typological thinking of Goldman's day was being replaced by the modern synthesis of paleontology, population genetics, and Darwinian evolution that emphasized variation within a species. Poole's (1940) paper apparently settled the *magister/pennsylvanica* question, as the synonymy of the two forms has not seriously been questioned since, and the designation "*N. pennsylvanica*" has dropped out of use.

Mid-1900 s: The Sinking of *N. magister*

While increasing numbers of specimens had done away with a false distinction between fossil and extant woodrats, the same trend soon threatened to eliminate *N. magister* as a species entirely. In the early years, woodrat taxonomists had compared Allegheny woodrats from the northern Appalachians with woodrats from the distant Gulf Coast and readily concluded that they were different. Goldman's (1910) mapped range for *N. floridana* was separated by hundreds of kilometers from the Allegheny woodrat's range. But as museum collections grew, closing the geographic gaps between the two, cline-minded biologists began proposing that the Allegheny woodrat was just a subspecies of the earlier described and very variable *N. floridana*.

The logic and techniques brought to bear on this question were new. Taxonomic works were beginning to be supported by more sophisticated statistical analysis of entire series of specimens. Rhoads' (1894) simplistic dichotomy of "cave rat" (*N. magister*) versus "wood rat" (*N. floridana*) would not accommodate a growing awareness of the variety of habitats used by eastern woodrats. Merriam's (1894) distinction of upland versus valley zones had similarly been eroded, as specimens

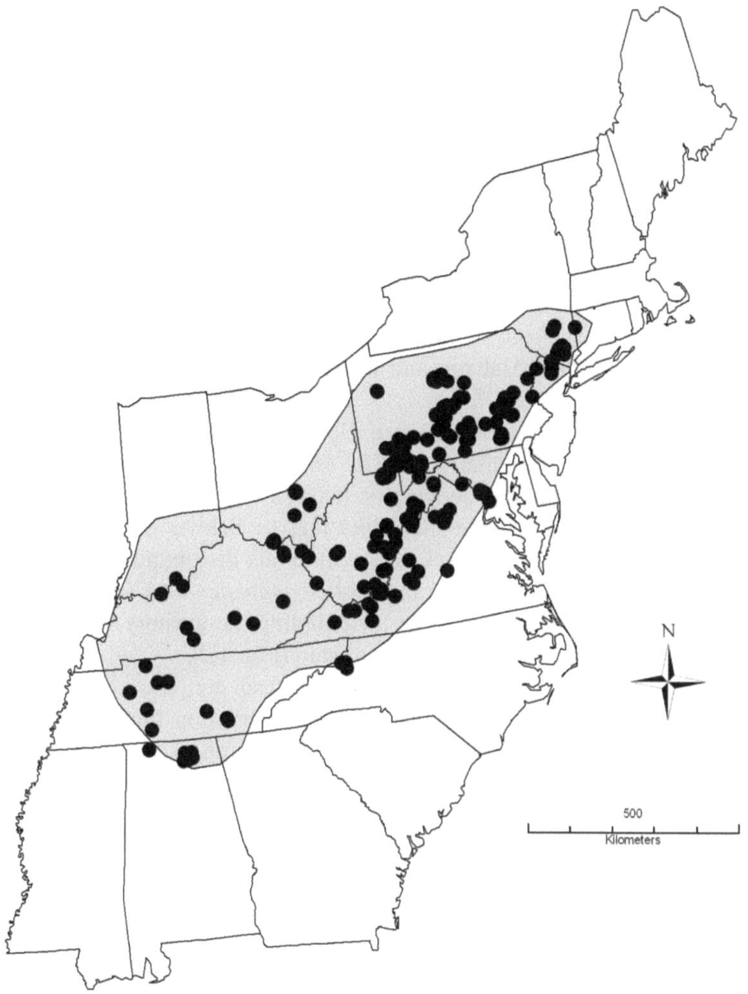

Fig. 1.2 Collecting localities for museum specimens of *Neotoma magister* by the time of Schwartz and Odum's (1957) revision (*closed symbols*), and complete range (*shaded*) as depicted in Hall (1981)

were collected in intermediate localities. By the mid-1950 s, museums held more than 700 Allegheny woodrat specimens collected from nearly 80 U.S. counties (Fig. 1.2).

Schwartz and Odum (1957) used these newer approaches to analyze taxonomic relationships among woodrats of the eastern United States. They examined variation in 12 morphological metrics taken from 224 specimens that they grouped by subspecies. Presenting their results in what at that time was a relatively novel graphic—a box-whisker plot showing means, standard errors, and ranges—they compared *N. magister* specimens with *N. floridana*, but divided the *N. magister* specimens into northern and southern samples. They argued that "northern" *N. magister* were indeed significantly larger than all *N. floridana*, but "southern" *magister* specimens (from Kentucky, Virginia, and southward) were intermediate in size and not significantly different from *N. floridana*. Schwartz and Odum (1957) claimed that earlier studies had exaggerated the size of *N. magister* by using a biased sample of primarily northern specimens. Clinal variation in size, a newly recognized phenomenon, could explain the bigger woodrats in the northern United States, they suggested, and it would be "more fitting to regard *magister* as a subspecies of *floridana*." They then reinforced this somewhat guarded wording with a new range map depicting just one eastern woodrat, *N. floridana*, with a continuous range from the Gulf Coast to Massachusetts. Within this range, the subspecies "*N. f. magister*" was represented from northern Alabama and central Tennessee northward (Schwartz and Odum 1957).

This revision received a boost when Hall and Kelson (1959) accepted their arrangement in the monumental work *The Mammals of North America*, the first comprehensive systematic and geographic listing of the American mammal fauna. Hall and Kelson's (1959) range map, like that of Schwartz and Odum (1957), showed *N. f. magister* as one arm of a continuous distribution for *N. floridana*. This map depicted a much larger range for *N. magister* than Goldman's (1910), citing museum specimens as the basis for doubling the territory in Tennessee and Kentucky and adding southern Indiana and Ohio (Fig. 1.2).

Relegating *N. magister* to subspecies status had a ripple effect through the secondary literature that strongly influenced conservation biology. Through the period of the 1960 s–1980 s, while conservation awareness for other species was growing, the Allegheny woodrat was almost universally treated as a subspecies of the "eastern woodrat" *N. floridana*. An important review of the biology of *N. floridana* (Wiley 1980) incorporated literature on the Allegheny woodrat into a composite description of biology that would make it difficult for subsequent researchers to tease them apart. State and regional mammal guides for the northeastern states written by mammalogists during this era listed the woodrat of the northeastern United States as *N. floridana* (Paradiso 1969, Gottschang 1981, Mumford and Whitaker 1982, Merritt 1987). The second edition of *The Mammals of North America* retained the subspecies designation (Hall 1981), while expanding the range further into Indiana and the mountains of Georgia. The most widely used popular mammal field guide (Burt and Grossenheider 1952) also used a range map that referred to all woodrats east of the Mississippi as "eastern woodrat." Although Birney (1976) did note that he was unable to hybridize "*N. f. magister*" in the laboratory with other *N. floridana* and suspected they were a separate species, his finding had little impact. Overall, the sources most readily available to management agencies, educators, and the general public during this period consistently presented the cave-and-fissure woodrat of the U.S. northeast as being a somewhat peripheral variant of the very common and widespread "eastern woodrat." In a context of competing issues and limited resources, there was little chance that the Allegheny woodrat would command much attention for conservation.

Molecular Approaches and the Re-establishment of *N. magister*

The development of new molecular methods in the early 1990 s created an opportunity to re-examine the taxonomic identity of the Allegheny woodrat. Hayes and Harrison (1992) conducted mitochondrial DNA analyses of 114 woodrats from 33 localities, including 49 *N. f. magister* woodrats from seven states, 50 individuals representing other subspecies of *N. floridana*, and 15 individuals of other *Neotoma* species. Analyses of mtDNA data produced cladograms that grouped all putative *magister* samples together, and demonstrated that *N. magister* was as divergent from *N. floridana* as other woodrat species were from each other (Hayes and Harrison 1992).

Conclusions based on molecular genetics were supported by morphological analyses. Hayes and Richmond (1993) studied skull and other morphological characteristics of 917 specimens representing 418 *N. f. magister* as well as the other six *N. floridana* subspecies. A principal components analysis based on morphological characteristics demonstrated almost complete separation of *magister* from *floridana* (Hayes and Richmond 1993). In fact, one cranial characteristic, the presence of a maxillovomerine notch, was virtually foolproof in distinguishing the two groups. Collecting localities of notch-bearing woodrats closely conformed to the previous descriptions of *magister* range, with "the single exception" of three woodrats collected at Muscle Shoals, Alabama, "just south of the Tennessee River" and outside *magister*'s range. However, historic maps of that locality before river impoundment indicate that Hayes and Richmond (1993) were probably mistaken in this "exception" and that the specimens in question were more likely collected on the north shore of the Tennessee, consistent with Hall's (1981) *magister* boundary (C. Ludwig pers. comm.).

The compelling combination of molecular and morphological evidence from the largest and most geographically comprehensive samples studied to that time prompted Hayes and Richmond (1993) to propose resurrection of *N. magister* as a species. Subsequent genetic analyses by Planz et al. (1996) supported this distinction in a context of the entire *N. floridana* species group. As further confirmation, Ray (2000) found a distinct karyotypic form and a trustworthy mitochondrial DNA marker to differentiate *N. magister* from *N. floridana haematoreia* in the southern Appalachians, with preliminary evidence that the same character could be used to distinguish *N. magister* from *N. floridana illinoensis* on the western border. The only evidence of introgression between *magister* and *floridana* was in two ambiguous specimens from Burke County, North Carolina (Ray 2000).

In light of the mounting evidence, *N. magister* was listed as a separate species in the Revised Checklist of North American Mammals (Jones et al. 1997, Baker et al. 2003). Also, it has been recognized in a world list of species (Wilson and Reeder 2005), profiled in a review of its biology (Castleberry et al. 2006), and was recently recognized in the Natural Heritage database NatureServe (2006). However, more than a decade after publication of the first molecular studies, many museums and websites still list this species as *N. floridana* (or even *N. pennsylvanica*), and much of the older literature is still in use. This lack of consistency makes it difficult to coordinate information and effort, as nonspecialists may not be aware that they are dealing with the same species under different names.

Determining Current Status

This volume characterizes the Allegheny woodrat as a species in decline, but we do not know, nor are we ever likely to determine, when and where its populations began to diminish. It is possible, however, and perhaps more instructive, to review the history of how the decline of this species was detected and perceived by biologists and the public. Documenting the decline of a species is complicated. The World Conservation Union (IUCN 2006) specifies criteria to determine conservation status of species worldwide. For example, the IUCN asks evaluators to quantify the geographic range of occupied habitat, estimate population size, and calculate the rate of decline over the past decade. In other words, it matters not just that a species has disappeared from places it once occupied, but also that it has such a trajectory of decline or has reached such a reduced condition that it appears doomed within a short period without intervention. All of the IUCN criteria are at least semiquantitative and demand supporting data.

Species status determination among the various U.S. states is considerably less standardized. The procedures always involve scientific data, but may also rely on expert opinion and political concerns. Supporting data may range from detailed long-term demographic studies to occasional questionnaires sent to conservation officers. In the absence of real data, newsletter or word-of-mouth accounts of what is happening in other states may have an influence. All of these methods have figured in the case of the Allegheny woodrat.

Several features of Allegheny woodrat biology make perceiving and documenting a population decline difficult (Mengak et al. Chapter 7). Because *N. magister* inhabits rock outcrops, cliffs, and caves that are often inaccessible, the habitat is patchily distributed and poorly mapped. Furthermore, woodrats are seldom observed directly, and trapping is laborious, expensive, and potentially disruptive. The lack of historic baseline data is also a problem. State-specific and regional mammal guides have often depicted distributional ranges in vague terms, especially for species perceived as common. Museum records and some published literature show historic woodrat sites, but location of many sites was not recorded precisely, so resurveying them involves guesswork. Even those museum records that are potentially valuable are frequently inaccessible. In addition, there is virtually no record of appropriate habitat that was searched in the past but found vacant. It is thus difficult to determine what fraction of appropriate habitat was historically occupied, or whether some current populations might even represent a range expansion.

A particularly complex issue that applies to the Allegheny woodrat is detecting a population decline in a metapopulation complex—a set of subpopulations that are linked by dispersal. The very nature of a metapopulation implies that even in a self-maintaining complex, individual sites will "wink out" from time to time and later be recolonized from nearby sites (Krohne 1997). Simply noting that woodrats are absent from a site previously populated is therefore not definite evidence of decline. Ideally, the entire metapopulation area needs to be surveyed repeatedly, and this has seldom been done or even considered.

Biologists have, however, been able to take advantage of one fortunate aspect of woodrat biology in detecting decline. Even if not directly observed, woodrats leave conspicuous, lasting sign of their occupation. Fecal piles ("latrines") are characteristic, winter food caches may be visible at rock surface, and plant-fiber nests may be seen in caves (Poole 1940). Experienced observers can distinguish fresh sign from sign produced in a previous season, so old sign can be used to detect former occupation. Using the evidence from woodrat sign, an observer can, in a single survey visit, determine with reasonable accuracy that a site was formerly occupied (old sign) but is now vacant (no fresh sign).

With these considerations in mind, we can examine how the Allegheny woodrat came to be recognized as a species of concern early on in some places, and only much later, if at all, in others:

Pre-Colonial range contraction—As early as 1940, Poole used woodrat bones from archaeological excavations in Pennsylvania to argue that the species' range before European settlement must have been considerably broader (Poole 1940). Similarly, Richards (1972, 1987) catalogued woodrat bones in numerous Indiana caves north of the species' current range. However, noting no "European materials" associated with these woodrat middens, Richards (1987) interpreted the range contraction as occurring prior to European settlement. Newcombe (1930) did claim that in pioneer days, woodrats "frequently came into the houses of settlers" and so must have been more common then, but until the 1970 s, most authorities considered woodrats abundant in the northeastern United States. For example, in Pennsylvania's statewide mammal survey in the late 1940 s, surveyors did not bother to trap for woodrats in some regions but merely noted that "every mountain talus slope and block area examined by us held woodrats" (Gifford and Whitebread 1951). The idea of a prehistoric larger range for the woodrat was consistent with findings for fossil bison, tapir, horse, and ground sloth, and did not draw the attention of conservationists.

New York—Various records through the 1970 s showed woodrats to be found in most of the available habitat in the Hudson Highlands and Shawagunk Ridge in counties west of the Hudson River. No alarm was raised until their disappearance was noted at the Mohonk Preserve in Ulster County around 1977. This report led to new surveys of historic habitat, in which only five sites were found to be occupied (Hicks 2005). The last known New York woodrats were removed in 1987 for captive propagation. At that time, the Allegheny woodrat was thought extirpated in New York. More recent sightings may represent migrants from New Jersey rather than a breeding population, but the species is now officially categorized as endangered in New York (Hicks 2005).

Connecticut and Massachusetts—Only one confirmed Allegheny woodrat record exists from Connecticut, a specimen collected prior to 1932 along the New York border (Goodwin 1932). The lone record from Massachusetts is a specimen in the museum of the University of Kansas that was collected in Berkshire County (N. Slade pers. comm.). The existence of actual populations has never been established in either state. Connecticut lists the woodrat as a species of concern, "believed extirpated," with no plan for reintroduction (Connecticut Department of Environmental Protection 2004).

New Jersey—Prior to 1960, woodrats had been collected from several locations in the four northernmost New Jersey counties. In the early 1980 s, concerned by reports of decline in New York, the New Jersey Division of Fish and Wildlife surveyed three historic woodrat sites but found no woodrats or recent sign there (Valent 2003). Searches of other appropriate habitat eventually located two previously unknown populations, one of which subsequently became extirpated. New Jersey designated the Allegheny woodrat an endangered species in 1991 based on these results (Valent 2003). The sole remaining woodrat population, along the Hudson Palisades, has been monitored with live-trapping periodically since 1987, and it has maintained a stable population size in recent years between about 20 and 30 (New Jersey Department of Environmental Protection 2005).

Pennsylvania—Biologists conducting bat surveys in the 1970 s noted old, but not fresh, woodrat sign in a number of Pennsylvania caves. This observation led to a systematic live-trapping investigation by Hall (1988) at sites where museum specimens of woodrats had been collected in statewide mammal surveys conducted around 1950. Of 54 historic sites trapped, 21 no longer supported woodrat populations by the late 1980 s (Hall 1988). Hall's alert led to the nomination of the Allegheny woodrat as a threatened species in Pennsylvania in 1981 (C. Butchkoski pers. comm.). Since that time, the Pennsylvania Game Commission has allocated annual effort to building a database of occurrence reports and field surveys to determine the distribution and abundance of woodrats in the state (Butchkoski 2006). These studies have documented recent disappearance of woodrats from much of the northern and eastern extent of the species' distribution in Pennsylvania, and some more central areas as well (Butchkoski 2006).

Maryland—Although the first Maryland woodrat specimens were taken on the Potomac River just outside the District of Columbia in 1921, Paradiso (1969) interpreted this part of the range as a quite recent expansion from Virginia. He implied that the mountains of western Maryland were the species' native range, and deemed woodrats "abundant" in Allegany County in western Maryland in 1961 (Paradiso 1969). There was little systematic effort to determine distribution or abundance of the species through the mid-1980 s, but Feldhamer et al. (1984) recommended that the woodrat be designated threatened in Maryland "because of its affinity for rocky areas, especially caves" which were considered at risk from frequent disturbance. After woodrat declines were reported in northern states, surveys and monitoring in Maryland in the 1990 s showed woodrats disappearing or already gone from many of the sites containing signs of previous occupation, and the species was designated endangered in the state in 1994 on the basis of this decline (D. Feller pers. comm.).

West Virginia—From the results of a statewide mammal survey and other records, McKeever (1954) showed the range of the Allegheny woodrat as covering nearly all of montane West Virginia (the southern and eastern portions of the state).

A considerable amount of ecological and genetic research has been done in the core region of Allegheny woodrat distribution in West Virginia (Castleberry et al. 2006), where woodrats have been presumed common and secure. In response to declines noted elsewhere, some sites that were trapped in the 1990 s were trapped again in 2004, with no pattern of decline evident (Stihler 2004). However, in a detailed demographic study, Manjerovic (2004) and Wood (Chapter 3) documented a population decline at their study site. The state's Wildlife Conservation Plan portrays West Virginia as the species' "stronghold" (West Virginia Division of Natural Resources 2005), yet the woodrat is noted in that plan as a species of greatest concern in a number of habitats, and it is singled out for special attention with plans for monitoring, surveying, mapping, and assessment of threats.

Virginia—Before 1950, Bailey (1946) and Handley and Patton (1947) compiled specimen and other records for the Allegheny woodrat to determine its distribution in Virginia. They reported woodrats in some 18 counties through the Appalachian chain and along the Potomac River. No measures of abundance were reported, but woodrats were described as "common from the Blue Ridge westward wherever suitable habitat occurs" (Handley and Patton 1947). Trapping studies through the 1990 s have showed no remaining populations along the Potomac River border with Maryland (Mengak 1999). Elsewhere in Virginia, many historical collecting sites are now vacant of woodrats, but other active sites have been discovered, so no overall trend is clear (Mengak 1999). The state has awarded the Allegheny woodrat no special concern status, and mentions it only as Tier IV (relatively low priority) for attention in Virginia's Wildlife Conservation Strategy (Virginia Department of Game and Inland Fisheries 2005).

North Carolina—Only one record of the Allegheny woodrat was noted in North Carolina by Hall and Kelson (1959), and only two by Clark (1987), who emphasized the lack of knowledge of this species in the western mountains of North Carolina and gave its status only as "undetermined." Ray (2000) trapped specimens at additional localities in six counties, and suggested that the Allegheny woodrat was more abundant in North Carolina than previously recognized. The strong similarity between *N. magister* and *N. f. haematoreia* of the mountains farther southwest has contributed to the confusion about this species. However, Ray (2000) developed methods to distinguish them karyotypically (Ray and Webster 2002) and using multiplex PCR analysis of mitochondrial DNA (Ray et al. 2002). The status of North Carolina woodrat populations is currently unknown (S. Bosworth pers. comm.), but North Carolina's Wildlife Conservation Strategy lists *N. magister* as a "priority species" for periodic monitoring (North Carolina Wildlife Resources Commission 2005).

Georgia—Although Cross (1955) and Golley (1962) stated that the Allegheny woodrat could be found on the Lookout Plateau of northwestern Georgia, Hayes and Richmond (1993) and Ray (2000) classified specimens from this region as *N. f. haematoreia* without comment, and state agency publications do not appear to recognize *N. magister* as a member of the fauna of Georgia.

Alabama—Howell (1921) depicted the Allegheny woodrat's range as that part of the state north of the Tennessee River, which would include the northernmost tier of counties. This distribution is consistent with trapping results by Zambernardi

(1956). Although there appears to have been little if any population-level analysis on this species in Alabama, it has been given a Priority 2 status ("high" but not "highest" conservation concern) with a recommendation to survey bluff habitats to better determine its distribution and status (Alabama Department of Conservation and Natural Resources 2005).

Tennessee—Some range maps for the Allegheny woodrat show it spreads over most of the state of Tennessee (e.g., Hall 1981), but museum records indicate a more limited historical distribution in two roughly parallel north–south bands, corresponding to the elevated Cumberland Plateau in the east and the Highland Rim in the west. There is a dearth of field data to determine possible decline, but based on expert opinion by state mammalogists, Tennessee has designated the Allegheny woodrat a Tier 1 species for high-priority attention in its Wildlife Action Plan (Tennessee Wildlife Resources Agency 2005). A recent study found active woodrat populations at 11 of 15 reported or suspected sites in the Cumberland Plateau (Parker 2006). These will be useful monitoring sites for future analysis.

Kentucky—The range of the Allegheny woodrat in Kentucky was only poorly known by 1974, when Barbour and Davis (1974) showed it spottily distributed in 11 central and eastern counties. No statewide field surveys have been done since then, but a database of "all historical data" has been assembled for use by the Kentucky Department of Fish and Wildlife Resources. This database shows a much broader distribution—42 counties had records within the last 70 years, and four more had only historic records before 1930 (Thomas 2003). Woodrat distribution in Kentucky falls into two regions, the Cumberland Plateau in the east and a Highland Rim karst band west of the center of the state. Whether populations statewide have increased, decreased, or remained stable is not yet known. A five-year study in Mammoth Cave National Park showed no trend (Thomas 2003). Krupa et al. (2004) found active woodrat populations in Breathitt County but not in adjacent Knott County, despite similar habitat and old sign. They hypothesized either that the vacancies in Knott County might be a normal part of the dynamics of a metapopulation, or alternatively that a genuine extirpation had resulted from fragmented habitat. At this point there is no clear indication of decline, but Kentucky has designated the Allegheny woodrat as a priority species for some local monitoring in its Wildlife Conservation Strategy (Kentucky Department of Fish and Wildlife 2005).

Ohio—By 1930, woodrats had been collected in five counties in Ohio (Peneston et al. 1999). The Allegheny woodrat was on Ohio's original endangered species list created in 1974 (A. LeCount pers. comm.). A 1976 survey located active populations only in Adams County, along the Ohio River, but not in other historic sites (Peneston et al. 1999). Another statewide assessment of all historic sites in 1998 found only a remnant few of the Adams County sites active and concluded that the species had "declined dramatically since its presence was first documented in the 1920 s" (Peneston et al. 1999). A more intense survey of Adams County showed little change in the number of sites between 1986 and 2004, but suggested a probable overall decrease in woodrat abundance (Mollohan and LeCount 2004). Analysis of raccoon scats collected from extirpated woodrat sites in Hocking and Fairfield

Counties showed 26% of samples infected with potentially lethal raccoon roundworm, whereas none from Adams County were infected (Mollohan and LeCount 2004).

Indiana—Indiana declared the Allegheny woodrat a threatened species as early as 1984 (Cudmore 1985), apparently not on the basis of a known decline but because available records showed a very limited distribution in the state. An extensive live-trapping study was then conducted to gain an accurate picture of the species' distribution. Woodrats were found at 20 sites in three southern counties along the Ohio River, with old but recent sign at four more sites. From these findings, Cudmore (1985) estimated the statewide population as 781 and recommended that the woodrat be listed as endangered. Again, this recommendation was apparently based on the potential vulnerability of a species with restricted habitat and low total numbers, rather than on evidence of precipitous recent decline. Since 1990, however, intensive efforts in Indiana have documented a clear decline to critical population levels (Johnson 2002).

Overall pattern—In the late 1980 s a coalition of biologists, concerned by indications of woodrat decline in the northeast, attempted to determine the status of the Allegheny woodrat. To supplement the scanty literature, they enlisted expert opinion in each state of the species' range, and summarized the results in a petition to the U.S. Fish and Wildlife Service to list the Allegheny woodrat under the Endangered Species Act (A. Linzey pers. comm.). The petition was denied (A. Linzey pers. comm.) on the grounds that the populations of concern, from Connecticut to Maryland and Ohio, were not "isolated" from those farther south where no decline was claimed. Indeed, the decline of the Allegheny woodrat has frequently been characterized as proceeding from the northern and eastern range perimeters. However, this apparent pattern may be an artifact of higher vigilance in northern states, with less scrutiny farther south. It is becoming increasingly clear that population declines have occurred through central Pennsylvania, western Maryland, and parts of Virginia and West Virginia, so it appears that central areas are not invulnerable. Indeed, it is not at all certain that the Allegheny woodrat is secure in its southern range through North Carolina, Alabama, and Tennessee, as very little recent survey work has been done in those states.

Conservation and Management of Declining Species

Conservation biologists considering the Allegheny woodrat, like those concerned with the clouded leopard, have had to grapple simultaneously with two slippery questions. On the one hand, is the species declining, and how can we tell? And on the other, just what is a clouded leopard, or an Allegheny woodrat? The answer to the taxonomic question is crucial for interpreting the demographic one.

This review has been written as if the designation of the Allegheny woodrat as *N. magister*, and its distribution as mapped here, are the final word on this species, but this may not be the case. As genetic analyses become more refined, previously inclusive taxonomic entities are being divided into several species. Cryptic species of woodrats in western North America have recently been recognized this way

(Edwards et al. 2001). If the Allegheny woodrat is really declining in some parts of its range yet secure in others, the difference in status may be due to varying severity of threats to this species, but there also may be genetic differences in resistance to those threats that could matter to conservation. Attention to the species' genetic identity will also help in drawing appropriate stock for reintroductions. It will be important for managers to keep abreast of genetic studies and for researchers to help managers understand the implications of their findings.

The knotty taxonomic history of the Allegheny woodrat has hampered efforts to understand not only its conservation status but also its overall biology. Much of the secondary literature of the last decades has combined data for *N. magister* and *N. floridana*, which have very different habitat requirements and behavior, and likely have different demography and relationships with predators, parasites, diseases, and mast food supply. Understanding the Allegheny woodrat requires disentangling the information for this species and identifying the gaps in knowledge. Attempts to accomplish this are underway (Castleberry et al. 2006). To succeed, attention to geography and taxonomy will be crucial. As future studies are conducted, it will be important, for this and other species of concern, to specify clearly the geographic sites of field studies and monitoring efforts. GPS and GIS tools should be universally used to accomplish and communicate this information with precision. The simple shaded range maps of the past will not be adequate to define a species' range if that range is patchy and subject to change.

The new technologies of genetics and digital mapping will be the essential tools. At the same time, we need to remember that our historic understanding of the taxonomic identity of most species is still based on studies of morphology, especially of museum specimens. Moreover, museum collections have been a unique resource for setting the baseline to detect Allegheny woodrat decline; for many other species, specimen-based records from the past are the only reliable geographic history that is available. Museum collections and their records are an irreplaceable resource that must be guarded for the future, and these data must be made more accessible.

As our understanding of population dynamics becomes more sophisticated, we need to define more carefully the criteria that we use to measure population changes. Spot checks of presence or absence will not be adequate. Where wildlife species contend with increasingly fragmented and disturbed habitat, we will need better definitions of "population" and "decline" to detect what is really taking place.

Finally, the disjointed state-by-state attention (or lack of it) that has been paid to the Allegheny woodrat is, unfortunately, not an isolated case. In preparing this review, I found repeatedly that experts in one state are often completely uninformed about what efforts or findings are taking place elsewhere. The federally mandated state Comprehensive Wildlife Conservation Strategy reports (also called Wildlife Action Plans), first produced in 2005, offer the first relatively standardized effort to communicate the status and projected future of wildlife in each state. These reports should be a first stop for an overview of any species of concern throughout its range, and future versions of these reports should be written with better inter-state coordination of knowledge and resources in mind.

References

Alabama Department of Conservation and Natural Resources. 2005. Alabama's comprehensive wildlife conservation strategy. Alabama Department of Conservation and Natural Resources, Montgomery, Alabama.

Allen, J. A. 1894. Remarks on a second collection of mammals from New Brunswick, and on the rediscovery of the genus *Neotoma* in New York State. Bulletin of the American Museum of Natural History 6:359–364.

Bailey, J. W. 1946. The Mammals of Virginia. Williams Printing Company, Richmond, Virginia.

Baird, S. F. 1857. Reports of Explorations and Surveys to Ascertain the Most Practicable and Economical Route for a Railroad from the Mississippi River to the Pacific Ocean. Volume 8, Part 1. Beverley Tucker, Washington, D.C.

Baker, R. J., Bradley, L. C., Bradley, R. D., Dragoo, J. W., Engstrom, M. D., Hoffman, R. S., Jones, C. A., Reid, F., Rice, D. W., and Jones, C. 2003. Revised checklist of mammals of North America north of Mexico, 2003. Occasional Papers Museum of Texas Tech University 229:1–23.

Barbour, R. W., and Davis, W. H. 1974. Mammals of Kentucky. University of Kentucky Press, Lexington, Kentucky.

Birney, E. C. 1976. An assessment of relationships and effects of interbreeding among woodrats of the *Neotoma floridana* species-group. Journal of Mammalogy 57:103–132.

Buckley-Beason, V., Johnson, W. E., Nash, W. G., Stanyon, R., Menninger, J. C., Driscoll, C. A., Howard, J., Bush, M., Page, J. E., Roelke, M. E., Stone, G., Martelli, P. P., Wen, C., Ling, L., Duraisingam, R. K., Lam, P. V., and O'Brien, S. J. 2006. Molecular evidence for species-level distinctions in clouded leopards. Current Biology 16:2371–2376.

Burt, W. H., and Grossenheider, R. P. 1952. A Field Guide to the Mammals (Peterson Field Guide Series). Riverside Press, Cambridge, Massachusetts.

Butchkoski, C. M. 2006. Allegheny woodrat research/management. Job Code 71801-05z, Appendix A, Allegheny woodrat habitat site survey code manual. Pennsylvania Game Commission, Harrisburg, Pennsylvania. Available http://www.pgc.state.pa.us/pgc/lib/pgc/reports/2006'wildlife/71801-05.pdf.

Castleberry, S. B., Mengak, M. T., and Ford, W. M. 2006. *Neotoma magister*. Mammalian Species 789:1–5.

Clark, M. K. 1987. Part I: A re-evaluation of the mammals. Endangered, threatened, and rare fauna of North Carolina. Occasional Papers of the North Carolina Biological Survey, 1987-3.

Connecticut Department of Environmental Protection. 2004. Connecticut's endangered, threatened, and special concern species 2004. Connecticut Department of Environmental Protection Hartford, Connecticut.

Cope, E. D. 1869. Synopsis of the extinct Mammalia of the cave formations in the United States. Proceedings of the American Philosophical Society 11:171–192.

Cross, W. H. 1955. Three *Neotoma* populations in Georgia and interpopulation mating in captivity. Journal of Mammalogy 36:572–573.

Cudmore, W. W. 1985. The present distribution and status of the eastern woodrat, *Neotoma floridana*, in Indiana. Proceedings of the Indiana Academy of Science 94:621–627.

Edwards, C. W., Fulhorst, C. F., and Bradley, R. D. 2001. Molecular phylogenetics of the *Neotoma albigula* species group: further evidence of a paraphyletic assemblage. Journal of Mammalogy 82:267–279.

Feldhamer, G. A., Gates, J. E., and Chapman, J. A. 1984. Rare, threatened, endangered and extirpated mammals from Maryland. *In* Threatened and Endangered Plants and Animals of Maryland, eds. A. W. Norden, D. C. Forester and G. H. Fenwick, pp. 395–438.National Heritage Program, Annapolis, Maryland.

Gifford, C. L., and Whitebread, R. W. 1951. Mammal survey of south central Pennsylvania. Final report to the Pittman-Robertson Project 38-R, Pennsylvania Game Commission, Harrisburg, Pennsylvania.

Goldman, E. A. 1910. Revision of the wood rats of the genus *Neotoma*. North American Fauna 31:1–124.
Golley, F. B. 1962. Mammals of Georgia: A Study of Their Distribution and Functional Role in the Ecosystem. University of Georgia Press, Athens, Georgia.
Goodwin, G. G. 1932. New records and some observations on Connecticut mammals. Journal of Mammalogy 13:36–40.
Gottschang, J. L. 1981. Guide to the Mammals of Ohio. Ohio State University Press, Columbus, Ohio.
Hafner, M. S., Gannon, W. L., Salazar-Bravo, J., and Alvarez-Castañeda, S. T. 1997. Mammal Collections in the Western Hemisphere. Allen Press, Lawrence, Kansas.
Hall, E. R. 1981. Mammals of North America. Second edition. John Wiley and Sons, New York.
Hall, E. R., and Kelson, K. R. 1959. Mammals of North America. Ronald Press, New York.
Hall, J. S. 1988. Survey of the woodrat in Pennsylvania. Contracts 878903 and 878919, Final report to the Pennsylvania Game Commission, Harrisburg, Pennsylvania.
Handley, C. O., and Patton, C. P. 1947. Wild Mammals of Virginia. Commonwealth of Virginia Commission, Richmond, Virginia.
Hayes, J. P., and Harrison, R. G. 1992. Variation in mitochondrial DNA and the biogeographic history of woodrats (*Neotoma*) of the eastern United States. Systematic Biology 4: 331–344.
Hayes, J. P., and Richmond, M. E. 1993. Clinal variation and morphology of woodrats (*Neotoma*) of the eastern United States. Journal of Mammalogy 74:204–216.
Hicks, A. 2005. Species group reports for mammals: Allegheny woodrat. New York State comprehensive wildlife conservation strategy. New York State Department of Environmental Conservation, Albany, New York.
Howell, A. H. 1921. Biological survey of Alabama. I. Physiography and life zones. II. The Mammals. North American Fauna 45:1–88.
IUCN. 2006. Guidelines for using the IUCN red list categories and criteria. Version 6.2. Standards and Petitions Working Group of the IUCN SSC Biodiversity Assessments Sub-Committee. Available http://app.iucn.org/webfiles/doc/SSC/RedList/RedListGuidelines.pdf.
Johnson, S. A. 2002. Reassessment of the Allegheny woodrat, *Neotoma magister*, in Indiana. Proceedings of the Indiana Academy of Science 1:56–66.
Jones, C., Hoffman, R. S., Rice, D. W., Baker, R. J., Engstrom, M. D., Bradley, R. D., Schmidly, D. J., and Jones, C. A. 1997. Revised checklist of North American mammals north of Mexico. Occasional Papers Museum of Texas Tech University 173:1–22.
Kentucky Department of Fish and Wildlife. 2005. Kentucky's wildlife conservation strategy 2005. Kentucky Department of Fish and Wildlife, Frankfort, Kentucky.
Krohne, D. T. 1997. Dynamics of metapopulations of small mammals. Journal of Mammalogy 78:1014–1026.
Krupa, J. J., Workman, J. W., Lloyd, C. M., Bertram, L. R., Horrall, A. D., Dick, D. K., Brewer, K. S., Valentine, A. M., Shaw, C. P., Clemons, C. M., Prater, C. A., Clemons, J. E., Jr., Campbell, N. J., Arnold, S. B., Jones, N. J., and Clark, A. M. 2004. Distribution and habitat use of the Allegheny woodrat (*Neotoma magister*) in a mixed-mesophytic forest in southeastern Kentucky. Journal of the Kentucky Academy of Science 65:33–38.
Mammal Networked Information System (MaNIS). 2007. University of California, Berkeley, California. Available http://manisnet.org/.
Manjerovic, M. E. 2004. Demographic and genetic structure of an Allegheny woodrat population in West Virginia. M.S. Thesis, West Virginia University, Morgantown, West Virginia.
McKeever, S. 1954. Ecology and distribution of the mammals of West Virginia. Ph.D. Dissertation, North Carolina State College, Raleigh, North Carolina.
Mengak, M. 1999. Woodrat studies. Job AB, Report to Project EW-2-4, Virginia Department of Game and Inland Fisheries, Richmond, Virginia.
Merriam, C. H. 1890. Results of a biological survey of the San Francisco Mountain region and the desert of the Little Colorado, Arizona. North American Fauna 3:1–136.

Merriam, C. H. 1894. A new superfamily of Murine rodents – the Neotominae – with description of a new genus and species and a synopsis of the known forms. Proceedings of the Academy of Natural Sciences of Philadelphia 1894:115–252.

Merritt, J. F. 1987. Guide to the Mammals of Pennsylvania. Carnegie Museum of Natural History, Pittsburgh, Pennsylvania.

Miller, G. S. 1940. Bones of mammals collected by Baird in Pennsylvanian caves. Journal of Mammalogy 21:319–321.

Mollohan, C. M., and LeCount, A. L. 2004. Status and distribution of the woodrat in Adams County, Ohio. Final report to the Ohio Division of Wildlife, Columbus, Ohio.

Mumford, R. E., and Whitaker, J. O., Jr. 1982. Mammals of Indiana. Indiana University Press, Bloomington, Indiana.

NatureServe. 2006. NatureServe explorer: an online encyclopedia of life. Version 6.1. NatureServe, Arlington, Virginia. Available http://www.natureserve.org/explorer.

New Jersey Department of Environmental Protection. 2005. Allegheny woodrat conservation. Mammal Conservation Job 1B, Federal Aid Project T-1-1, New Jersey Department of Environmental Protection, Trenton, New Jersey.

Newcombe, C. L. 1930. An ecological study of the Allegheny cliff rat. Journal of Mammalogy 11:204–211.

North Carolina Wildlife Resources Commission. 2005. North Carolina wildlife action plan. North Carolina Wildlife Resources Commission, Raleigh, North Carolina.

Paradiso, J. L. 1969. Mammals of Maryland. North American Fauna 66:1–193.

Parker, W. 2006. Immobilization of small mammals and occupancy, seasonal food habits, and parasites of Allegheny woodrats in the Cumberland Mountains of Tennessee. M.S. Thesis, University of Tennessee, Knoxville, Tennessee.

Peneston, W., Mollohan, C., Enterline, K., and LeCount, A. 1999. Status and distribution of the endangered woodrat in Ohio. Final report to the Ohio Biological Survey, Columbus, Ohio.

Planz, J. V., Zimmerman, E. G., Spradling, T. A., and Akins, D. R. 1996. Molecular phylogeny of the *Neotoma floridana* species group. Journal of Mammalogy 77:519–535.

Poole, E. L. 1940. The technical name of the Allegheny woodrat. Journal of Mammalogy 21: 316–318.

Ray, D. K. 2000. Phylogenetics and evolution of woodrats (Genus *Neotoma*) in the southern Appalachian Mountains. M.S. Thesis, University of North Carolina at Wilmington, Wilmington, North Carolina.

Ray, D. K., Freshwater, D. W., and Webster, W. D. 2002. Multiplex PCR identification of eastern and Allegheny woodrats. Journal of Wildlife Management 66:86–90.

Ray, D. K., and Webster, W. D. 2002. Karyotypic analysis of *Neotoma* (Rodentia: Muridae) from the eastern United States. Journal of the North Carolina Academy of Science 120:50–54.

Rhoads, S. N. 1894. A contribution to the life history of the Allegheny cave rat, *Neotoma magister* Baird. Proceedings of the Academy of Natural Sciences of Philadelphia 46:213–221.

Richards, R. L. 1972. The Woodrat in Indiana: recent fossils. Proceedings of the Indiana Academy of Science 81:370–375.

Richards, R. L. 1987. The Quaternary distribution of the eastern woodrat, *Neotoma floridana*, in Indiana. Proceedings of the Indiana Academy of Science 96:513–521.

Schwartz, A., and Odum, E. P. 1957. The Woodrats of the eastern United States. Journal of Mammalogy 38:197–206.

Stihler, C. 2004. Rare and endangered species update. West Virginia Wildlife Magazine Winter 2004:1–2.

Stone, W. 1893. Description of a new species of *Neotoma* from Pennsylvania. Proceedings of the Academy of Natural Sciences of Philadelphia 45:16–18.

Tennessee Wildlife Resources Agency. 2005. Tennessee's comprehensive wildlife conservation plan. Tennessee Wildlife Resources Agency, Nashville, Tennessee.

Thomas, S. C. 2003. Allegheny woodrat monitoring. Cooperative Agreement G5530020001, Final report to the Kentucky Department of Fish and Wildlife, Frankfort, Kentucky.

Valent, M. 2003. Allegheny woodrat, *Neotoma floridana magister*. *In* Endangered and Threatened Wildlife of New Jersey, eds. B. E. Beans and L. Niles, pp. 7–12. Rutgers University Press, New Brunswick, New Jersey.

Virginia Department of Game and Inland Fisheries. 2005. Virginia comprehensive wildlife conservation strategy. Virginia Department of Game and Inland Fisheries, Richmond, Virginia.

West Virginia Division of Natural Resources. 2005. West Virginia wildlife conservation action plan. West Virginia Division of Natural Resources, Morgantown, West Virginia.

Wiley, R. W. 1980. *Neotoma floridana*. Mammalian Species 139:1–7.

Wilson, D. E., and Reeder, D. M. (eds.). 2005. Mammal Species of the World. Johns Hopkins University Press, Baltimore.

Wilting, A., Fischer, F., Abu Bakkar, S., and Linsenmair, K. E. 2006. Clouded leopards, the secretive top carnivore of South-East Asian rainforests: their distribution, status and conservation needs in Sabah, Malaysia. BMC Ecology 6:16.

Zambernardi, J. 1956. Woodrats of the genus *Neotoma* in Alabama. M.S. Thesis, University of Alabama, Tuscaloosa, Alabama.

Chapter 2
Multiple Causes of the Allegheny Woodrat Decline: A Historical–Ecological Examination

Kathleen LoGiudice

Introduction

Wandering through the former haunts of the Allegheny woodrat (*Neotoma magister*) in the Shawangunk Mountains of New York State, one might easily imagine that this landscape is largely unchanged from what it was hundreds of years ago. Massive chestnut oaks (*Quercus prinus*) protrude from rocky hillsides and steep talus slopes conceal complex systems of fissures and galleries beneath. However, one need not look far to see signs of changing land uses all around this beautiful region. Even in areas seemingly untouched by human hands, the plant and animal communities of northeastern North America have undergone profound changes in the past 400 years—a fact not often considered when pondering the causes for a species' decline. And yet, a short moment of reflection brings to mind a plethora of important species that no longer exist in our landscape. The American chestnut (*Castanea dentata*) and the American elm (*Ulmus americana*) are no longer found in our forests. The eastern hemlock (*Tsuga canadensis*), American beech (*Fagus grandifolia*), and eastern dogwood (*Cornus florida*) are fast disappearing. The list of animal species once found in the northeast but now absent is also quite impressive, including wolves (*Canis lupus*), cougars (*Felis concolor*), passenger pigeons (*Ectopistes migratorius*), wolverines (*Gulo gulo*), bison (*Bison bison*), and lynx (*Lynx canadensis*).

Equally noteworthy is the reversal of devastating trends that once appeared unidirectional; deforestation gave way to reforestation which has given way to suburbanization. Some species appear to have disappeared from the northeast forever. Others, like white-tailed deer (*Odocoileus virginianus*), American turkeys (*Meleagris gallopavo*), black bears (*Ursus americanus*), and American beavers (*Castor canadensis*) declined only to rebound to levels once thought impossible. From a nadir in the 1850s, when Henry David Thoreau noted that the muskrat

K. LoGiudice
Department of Biological Sciences, Union College, Schenectady, NY 12308

J.D. Peles, J. Wright (eds.), *The Allegheny Woodrat*, DOI: 10.1007/978-0-387-36051-5_2,

(*Ondatra zibethicus*) was the largest animal in the Massachusetts countryside (Foster 2000), we have seen a steady rise in the numbers of increasingly bold, human-adapted species. Deer feast on our shrubberies, turkeys strut in our backyards, and countless other species benefit from the changes we have made to the landscape. In addition, non-native species have proliferated in our forests and cities. Both through natural range expansions and deliberate or accidental introductions, many species that were unknown in eastern North America even 50 years ago are now well established.

Historical changes in the plant and animal communities must form the backdrop when we consider the changing population dynamics of the Allegheny woodrat (*Neotoma magister*). Woodrats have experienced repeated changes in food resources, competitors, predators, parasites, and landscape characteristics. Thus, impacts of adverse factors like the loss of an important food resource might be compounded or even ameliorated by other concurrent changes in the competitive or predatory regime to which the woodrat was exposed, causing a temporal disconnect between the onset of the stressor and the population response.

This mosaic of interacting factors might explain why the reasons for the decline of the Allegheny woodrat remain enigmatic. None of the factors most commonly believed to cause extinction—overharvesting, habitat destruction, impacts of introduced species, or secondary extinction (Diamond 1989)—seem capable of explaining the widespread disappearance of the species. Woodrats are economically unimportant either as a resource or as a pest, so overkill cannot explain their demise. The talus and cliff faces that form woodrat habitat proper are generally intact and no introduced predator or competitor has directly threatened the species. The introduction of the chestnut blight (*Cryphonectria parasitica*) and subsequent elimination of the American chestnut has been suggested as a case of secondary extinction but the woodrat survived this loss, even in the extremes of its range, by at least 50 years (Campbell and Madden 1990). Nevertheless, the possible factors that explain the species' demise are all related to human actions. In this chapter, I will examine the decline of the Allegheny woodrat by focusing on the three hypotheses that have been suggested to explain it. I also propose a fourth hypothesis that incorporates the first three, but places them in their "historical-ecological" context (Foster 2002) and considers whether concurrent ecological changes may help to explain the strength of the response. I then present a hypothetical chronology of the events leading to the decline of the Allegheny woodrat.

Decline of the Allegheny Woodrat

As outlined in Chapter 1, the former range of the Allegheny woodrat extended from southern New York and Connecticut to northern Alabama. The current decline was first noticed in the late 1970s in Pennsylvania and New York, but there is some evidence of a long-term constriction of its range, particularly in the northern extremes. Fossil remains several thousand years old have been found in caves near Albany, NY, approximately 80 km north of the recently known range (Hicks 1989, Whitaker and Hamilton 1998). There is evidence that the range also has constricted since the Pleistocene in Ohio and Indiana (Cudmore 1983, Whitaker and Hamilton 1998). Thus, it is possible that the observed decline is a continuation and acceleration of a long-term trend occurring for thousands of years.

The older scientific literature contains some suggestions that woodrats are intolerant of coexistence with humans. This was first mentioned by Peter Kalm in 1759 (cited in Rhoads 1894). In this early account, the woodrat was described as living "far from mankind among the stones and caverns in the Blue Mountains." Rhoads (1903) stated that "any permanent inroads of civilization into their territory result in their speedy extermination." Several early 20th century accounts of the Allegheny woodrat mention a perceived decline in its numbers. Newcombe (1930) described the Allegheny woodrat as being far less common than it was in "former years" and, in contrast to earlier intimations that they avoid humans, reported that "in the early days it frequently came into the houses of settlers" and carried away dried fruits and other small objects. Poole (1940) commented that, despite being well adapted to predator avoidance, "the animal seems to have narrowed its range in recent times to the immediate vicinity of rock ledges, boulder tumbles, and caves." Nevertheless, the records of the New York State Museum show that the Allegheny woodrat survived, even in the northern-most part of its recent range, at least until the 1960s (Hicks 1989).

The historical–ecological approach has been extremely successful in interpreting forest vegetation patterns, exposing the highly dynamic nature of natural systems and helping to frame conservation issues (Eberhardt et al. 2003), but it has been less commonly used to investigate changes in wildlife populations. This may be due to the inherent difficulty in measuring population dynamics of animals. It is difficult enough to estimate the size of any animal population today, and to do this retroactively is even more challenging and speculative. However, some authors have succeeded in combining historical, scientific, and cultural sources with trapping records and fur prices to trace the dynamics of many highly visible and economically important animal species (McDonald and Harris 1999, Zielinski et al. 2001, Foster et al. 2002, Gompper and Hackett 2005).

No such historical records exist for the Allegheny woodrat, but these works provide the context in which the dynamics of less studied species can be examined. By taking into account the changing competitive and predatory regimes that might have magnified or minimized the impacts of other relevant perturbations, the important causative factors might be more easily identified and possibly remedied. Although it is not clear that we can save the Allegheny woodrat, its story teaches us to consider multiple hypotheses when investigating the decline of a species, to examine the possibility of additive, compensatory, synergistic and/or threshold effects, and to take the long-term perspective by being cognizant of how historical fluctuations in the biotic environment may have influenced the timing and pattern of decline.

Hypotheses to Explain the Decline

The three most common hypotheses to explain the decline of the Allegheny woodrat are: (1) decrease in food availability; (2) habitat fragmentation/disturbance; and (3) parasite mortality (Balcom and Yahner 1996).

Food Decline Hypothesis

The food decline hypothesis was suggested by Hall (1987), based upon his observation that the disappearance of Allegheny woodrat populations in Pennsylvania during the late 1970s was concurrent with intense gypsy moth (*Lymantria dispar*) infestations that had been affecting these regions since early in the decade (McManus and McIntyre 1981). Repeated defoliations can cause mast failure and oak mortality (Liebhold et al. 1994). The food decline hypothesis suggests that woodrat populations were responding to a decline in the availability of winter food in the form of acorns, an important component of the Allegheny woodrat diet (Castleberry et al. 2002). Hall posited that the reduction in acorn availability was affecting woodrats in two ways: (1) by forcing increased movements during winter, thereby increasing exposure to weather and predation; and (2) by reducing reproductive output as animals entered the breeding season with depleted fat reserves (Hall 1987, Balcom and Yahner 1996).

There are two lines of evidence to suggest that woodrats need good winter nutrition to launch a successful reproductive effort. The first involves correlative data from various woodrat species in the field; secondly, evidence from reproductive physiology experiments on eastern woodrats (*N. floridana*). Several investigators have observed that woodrat populations have increased after the production of an abundant acorn crop or collapsed after a complete mast failure. Murphy (1952) observed a large increase in an eastern woodrat population during two years of abundant acorn crops (1949 and 1950). Anecdotally, a record high population of Allegheny woodrats was recorded in New Jersey in the year following an exceptional mast year (J. Sciascia and K. LoGiudice unpublished data), and a precipitous crash of a thriving Allegheny woodrat population occurred immediately after a mast failure in Maryland (D. Feller pers. comm.).

A comparison of two experimental reintroductions of Allegheny woodrats also indicates that good winter nutrition results in a strong reproductive effort. In both studies, animals were introduced in the late summer after the breeding season had ended. The first reintroduction (McGowan 1993) was conducted in an exceptional oak mast year, whereas the second reintroduction (LoGiudice 2003) was carried out during a year with a moderate acorn crop. In the following breeding season, average recruitment in the first study (4.8 young weaned per female) was more than twice as great as the second study (2.3). This difference would perhaps have been more extreme if females in the second study had not been provided with supplemental food (LoGiudice 2003). In that study, two females survived the winter and the entire breeding season—one supplementally fed, and the other a control. In this small sample, the difference in reproductive success was striking. The food-supplemented animal weaned at least five young in three litters and the control animal weaned only one offspring. In both reintroductions, there appears to have been enough food because all animals gained weight throughout the fall (K. LoGiudice unpublished data, E. McGowan and A. Hicks unpublished data). However, it is possible that the smaller acorn crop prevented animals from caching enough food to last through the winter, resulting in females entering the breeding season with less than optimal fat reserves.

Although little has been published on the reproductive physiology of the Allegheny woodrat, the reproductive energetics of the eastern woodrat have been well studied. Both species share similar reproductive cycles, which include small litters, slow postnatal development, and an unusually long lactation period (Poole 1940, McClure and Randolph 1980, Mengak 2002). In a series of experiments, McClure and colleagues compared reproductive energetics of the eastern woodrat to the hispid cotton rat (*Sigmodon hispidus*; McClure and Randolph 1980, McClure 1987). They found that eastern woodrats (a "slow" reproducer) require a larger energetic output per pup than do cotton rats (a "fast" reproducer), in part because of the slow development of the pups (McClure and Randolph 1980). When food was severely restricted during lactation (McClure 1987), cotton rats were able to successfully wean 75% of their young, whereas only 47% of the woodrat young survived to weaning. The author noted that many more woodrat pups would have died if she had not removed woodrat mothers from the experiment when their weight decreased below a certain threshold. When these experiments were repeated (Sikes 1995), somewhat better woodrat pup survival was reported (68%), but there was still a high cost to maternal malnutrition. In addition, under conditions of food restriction, woodrats depend on stored fat more heavily than cotton rats, indicating that reproductive success in woodrats is more heavily influenced by maternal prebreeding condition (McClure 1987). McClure hypothesized that the slow reproductive cycle of woodrats is an adaptive response to limited food resources and a stable, predictable food supply. It is possible that, in the case of the Allegheny woodrat, this stable food supply was the American chestnut (E. McGowan pers. comm.), with which it shares its entire range.

Many authors (Balcom and Yahner 1996, Wright and Kirkland 2000, LoGiudice 2003, 2006, Ford et al. 2006) have expanded the food decline hypothesis to include the impact of loss of the American chestnut by the chestnut blight. This parasitic fungus eliminated chestnuts from the range of the woodrat beginning in about 1910 and continuing through the 1930s (Woods and Shanks 1959, Day and Monk 1974, Campbell and Madden 1990). Before its destruction, the American chestnut was the dominant tree in the Appalachian forests, both in abundance (comprising about 25% of all woody stems) and in size (growing to nearly 4 m in diameter; Elias 1980, Brewer 1995). It is likely that the loss of the food resources provided by the American chestnut had a significant impact on Allegheny woodrat populations.

Unfortunately, the historical importance of American chestnuts in the Allegheny woodrat diet can only be inferred since chestnuts are rarely mentioned by early authors. As most woodrat research was done after the destruction of mature American chestnuts, this is not necessarily indicative of the former importance of chestnuts in the woodrat diet. Only Poole (1940) reported direct evidence of chestnuts in the diet—chestnut meat in analyses of stomach contents conducted over a number of years starting in 1922. The blight had killed more than 50% of the mature chestnut trees in Poole's study site in Berks County, Pennsylvania by 1922 and probably quite a bit earlier since more than 50% of the trees in the county were reported to be infected by 1912 (Pennsylvania Chestnut Blight Conference 1912, Campbell and Madden 1990). Thus, evidence of chestnuts in the diet may imply that chestnut meat was a particularly important resource. Today the populations studied by Poole in Berks County are extirpated (Pennsylvania Game Commission 1995).

It is well established that populations of other North American rodents, such as the white-footed mouse (*Peromyscus leucopus*), cycle with oak masting events. Populations reach its peak during summer after a heavy mast year and crash after a mast failure (Wolff 1996, Jones et al. 1998). However, mice, voles, and other cricetids can produce multiple large litters of young per year (Whitaker and Hamilton 1998), whereas woodrats rarely produce more than two litters of two to four young annually (Whitaker and Hamilton 1998, Mengak 2002). Furthermore, other cricetids reach sexual maturity quickly, producing several generations per year, whereas woodrats do not breed during their first year (McClure and Randolph 1980). The comparatively slow reproductive rate of woodrats would appear to make it more crucial to maintain steady, predictable reproductive output from year-to-year.

Stable interannual reproductive rates may not have been possible given the changes to food resources experienced by Allegheny woodrats in the past 80 years. Woodrats can apparently sustain themselves on a high fiber diet of foliage and browse when necessary. However, seeds and fruits are preferred when available (Rhoads 1903, Newcombe 1930, Poole 1940, Castleberry et al. 2002, Castleberry and Castleberry Chapter 6). It is possible that the Allegheny woodrat can survive during the winter on a poor quality fibrous diet, but requires a higher quality diet supplemented with nuts and seeds for successful reproduction, as has been demonstrated for several western *Neotoma* species (Justice and Smith 1992). If so, interannual reproductive output may have become more erratic as woodrats adapted to the boom and bust masting cycle of oaks (McShea and Schwede 1992) from the reliable annual crops of the American chestnut (Brewer 1995, Anagnostakis 2000). This could result in large fluctuation in the size of populations and exaggerate the effects of stochastic mortality sources.

Beyond the loss of the chestnut and the invasion of the gypsy moth, changing forest conditions have converged to challenge the dominance of oak species in the eastern forests and reduce the availability of hard mast. Altered forest management regimes and suppression of fire have favored faster growing species such as maples (*Acer* spp.) and birch (*Betula* spp.) at the expense of the fire tolerant oaks (Spurr and Barnes 1980, Shotola et al. 1992, Healy et al. 1997). Overabundant white-tailed deer populations consume acorns and young oak trees, further suppressing oak recovery (McCabe and McCabe 1997). All these forces have likely reduced the share of nutritious nuts available to Allegheny woodrats.

Habitat Fragmentation/Destruction Hypothesis

A second hypothesis states that habitat destruction and/or fragmentation has caused the extirpation of Allegheny woodrat populations (Balcom and Yahner 1996, Ford et al. 2006). Allegheny woodrats are habitat specialists, found primarily in rocky outcrops, caves, cliff faces and talus slopes (Whitaker and Hamilton 1998). There is little evidence for wide-scale destruction of these habitats in the northern portions of the woodrat range where the species decline is most pronounced. Most, if not all, historic woodrat sites in New York, New Jersey, Pennsylvania, and Maryland have been visited in the past 20 years by state conservation agencies and there have been very few reports of loss of rocky habitat (D. Feller pers. comm., Hassinger and Butchkoski 1999, A. Hicks pers. comm., J. Sciascia pers. comm.). Woodrat habitat in the southern part of the range where veins of coal underlie the cliffs may have undergone more direct disturbance. Coal mining completely destroys habitat, but as this occurs primarily on private lands, the extent to which mining has impacted woodrat populations is largely unknown (Chamblin et al. 2004, Ford et al. 2006). Chamblin et al. (2004) found that there was some use of reclaimed mining lands by woodrats, although the extent to which these habitats can be permanently occupied is still an open question. However, throughout the range of the Allegheny woodrat, forests surrounding the talus have had varying degrees of disturbance and there is some evidence that fragmentation may be adversely affecting this species. The patchy distribution of Allegheny woodrat populations could magnify the effects of fragmentation if major dispersal routes were blocked.

Several authors have examined sets of extant and extirpated woodrat habitat sites and compared the probability of occupation by woodrats with multiscale habitat and landscape variables. In Pennsylvania, Balcom and Yahner (1996) found that the probability of colony occupancy increased with distance from the forest edge, with the detrimental effects of the edge evident for up to 2 km. They went on to suggest that woodrats might be impacted by the larger populations of predators such as great horned owls (*Bubo virginianus*) that are often associated with agricultural development adjacent to the sites.

Ford et al. (2006), working in Maryland, Virginia, and West Virginia, used a spatially explicit model selection approach to compare the fit of several multiple logistic regression models to woodrat presence–absence data. Independent variables included distance to water, roads and railways, percent forest, agricultural, mining and pastoral cover, and most presciently, distance to nearest occupied rock outcrop. By far, the most important predictor of site occupancy was distance to the next nearest occupied site. Occupied sites were an average of 5 km from the next nearest active site; extirpated sites were more than twice as far on average (11.1 km). The most parsimonious model included the nearest neighbor distance, physiographic subprovince (Allegheny Plateau in WV, Ridge and Valley in WV/VA, and the Blue Ridge/Piedmont in VA/MD), and position relative to the Potomac River (north or south). Once distance to nearest occupied site was incorporated, probability of extirpation increased from west to east and was higher north of the Potomac. Also, there was some support for a second model that added percent forest area within 1 km of the colony site. This pattern of north to south and east to west extinction seems consistent with patterns in the more northern parts of the range. Ford et al. (2006) posited that a longer history of human development in the north and east may explain at least part of the pattern.

Parasite Hypothesis

Increased mortality caused by the parasitic raccoon roundworm (*Baylisascaris procyonis*; McGowan 1993, Birch et al. 1994, Balcom and Yahner 1996, LoGiudice 2003, 2006) is the third hypothesis for the decline of the Allegheny woodrat. The raccoon (*Procyon lotor*) serves as the definitive host for this intestinal nematode, but the parasite can also invade many species of intermediate hosts during its complex life cycle. The mature worms inhabit the small intestine of a raccoon and the eggs are passed in its feces. The sticky eggs are ingested by birds and mammals as they forage for seeds and undigested material in raccoon latrines, or are picked up passively and groomed off the feathers or fur (Page et al. 1999). Once ingested, larvae embark on an extended migration through the host tissues with some reaching the central nervous system where they almost always result in the death of the host, either directly or by causing behavioral changes that make the animal more vulnerable to predation (ataxia, circling, etc.). The cycle of transmission is completed when the intermediate host is consumed by a raccoon. Raccoon roundworm has been shown to cause pathology in more than 50 species of avian and mammalian intermediate hosts infected naturally or experimentally (Kazacos 2001).

Developed from a study conducted by the New York Department of Environmental Conservation (DEC) in 1991 (McGowan 1993), the parasite hypothesis is supported by an experimental trial conducted in New York and New Jersey (LoGiudice 2003). Both were translocation studies and in each case, woodrats introduced into sites heavily contaminated with raccoon roundworm eggs survived for significantly less time than those introduced into less contaminated sites. The DEC study was aimed at determining causes of mortality and attempted to recover dead and moribund animals from the talus. Ten of the 11 animals recovered for necropsy had *B. procyonis* larvae in the brain, and an examination of the release sites revealed extensive contamination with infected raccoon feces (McGowan 1993).

In a subsequent investigation conducted in 1996–1997, LoGiudice (2003) tested the parasite hypothesis by releasing woodrats into four historical sites in New York and New Jersey. Two of the release sites were heavily used by raccoons and contained many infected feces (high exposure sites) and two had lower levels of contamination (low exposure sites). LoGiudice found that animals survived significantly longer in the low exposure sites than in the high exposure sites (33.6 vs 15.5 weeks, respectively), supporting the parasite hypothesis. Allegheny woodrat mortality caused by *B. procyonis* also has been confirmed in Pennsylvania (J. Wright pers. comm.) and Indiana (Hudson et al. 2006), and infected raccoon feces have been found in rock outcrops occupied by woodrats in Maryland (D. Feller pers. comm.).

The behavior of collecting and storing a broad array of items, both edible and nonedible, seems to put woodrats at higher risk to raccoon roundworm than other small mammals. They appear to be attracted to the feces of raccoons and other animals (Rhoads 1894, Poole 1940, McGowan 1993, LoGiudice 2001, 2003, Hudson et al. 2006), collecting and caching whole feces regardless of seed content. In contrast, white-footed mice show some aversion to feces and remove seeds from scat immediately (LoGiudice 2001). Since roundworm eggs are harmless until embryonated, a process that takes up to 30 days, the caching of feces by woodrats puts them at a considerably higher risk to roundworm mortality than mice. Under the conditions in caves and talus passages, the eggs can remain infective for years (Kazacos 2001), prolonging the period of possible fatal infection. Furthermore, woodrats sequentially use dens for many generations (Newcombe 1930). Thus, a contaminated den could become a "parasite trap" that could kill many successive residents, as was observed by both McGowan (1993) and LoGiudice (2003). Allegheny woodrats have little or no resistance to *B. procyonis* infection. When infected with a standard dose of infective *B. procyonis* eggs, woodrats were significantly more likely than white-footed mice to develop fatal neurological disease (K. Kazacos pers. comm.).

Raccoon roundworm seems certain to have played a role in the disappearance of the Allegheny woodrat from New York, New Jersey, and probably Connecticut. There is evidence that raccoon population densities increased dramatically during the mid-20th century due to loss of predators, changes in land use, and relaxation of hunting and trapping pressures (Balcom and Yahner 1996, LoGiudice 2003, Prange and Gehrt 2004). This is likely to have increased woodrat exposure to raccoon roundworm. How common this mortality source is in other parts of the species' range is not clear. Owen et al. (2004) performed fecal analyses on raccoons trapped in woodrat habitat in West Virginia and found that *B. procyonis* was completely absent. Although the sample size was somewhat low ($n = 9$) for animals collected during the fall period of highest egg shedding and the detection technique is not the most effective (fecal analysis is the least effective method of detecting infection compared to latrine analysis and necropsy; Page et al. 2005), it is likely that if *B. procyonis* was present even at low prevalence, it would have been detected in at least one of the animals. In general, although raccoon activity is commonly detected in and around woodrat sites in many parts of the range (Owen et al. 2004, Hudson et al. 2006), *B. procyonis* eggs have only sporadically been found in feces. This may be partly due to the time of sampling. Raccoon roundworm shows strong seasonality, with peaks in egg shedding in fall and winter (Page et al. 2005); so sampling should be conducted with this in mind. In addition, the window of time in which the effects of *B. procyonis* might be detected in a woodrat population is exceedingly short. Cached feces are infrequently found by investigators, and the parasite appears to be so debilitating in woodrats that the probability of capturing an infected animal is low.

Circumstantial evidence in the woodrat monitoring data from Indiana (Johnson 2002, Hudson et al. 2006) suggests that raccoon roundworm might be an important mortality source in this state. Each of the three monitoring sites with the largest declines in individuals trapped between 2002 and 2005 was reported to have extensive raccoon sign during the 2005 trapping session. In addition, a fourth site at which a *B. procyonis* infected woodrat was captured in 1996 declined from 20 animals in 1996 to 0 in 2002. In three of the four sites experiencing the smallest declines between 2002 and 2005, the authors noted the absence of raccoon activity. With high roundworm prevalence previously reported in parts of Indiana (Jacobson et al. 1982, Kazacos 2001, Page et al. 2005), it is unlikely that *B. procyonis* is not at least periodically present in woodrat sites. A project currently underway is poised to shed some light on the importance of roundworm in Indiana (T. Smyser pers. comm.) and the regular monitoring suggested by the Indiana Department of Natural Resources (Hudson et al. 2006) is likely to help determine the possible impact of this parasite. One final piece of circumstantial evidence comes from the New Jersey woodrat monitoring project, which has documented a steady increase in the last remaining woodrat population (J. Sciascia, M. Valent and K. LoGiudice unpublished data; Fig. 2.1). This increase began several years after raccoon rabies first appeared in the northeast, presumably reducing what had been extremely high raccoon densities.

The importance of *B. procyonis* in the more southern part of the Allegheny woodrat range is very unclear. It has been long believed that prevalence of the parasite declined dramatically with latitude, but recent studies have revealed medium to high prevalences in Georgia (Eberhard et al. 2003) and Texas (Long et al. 2006). Although there are still unanswered questions about the seeming sudden impact of this native parasite, it appears to be common enough that wildlife managers should investigate raccoon activity in woodrat sites.

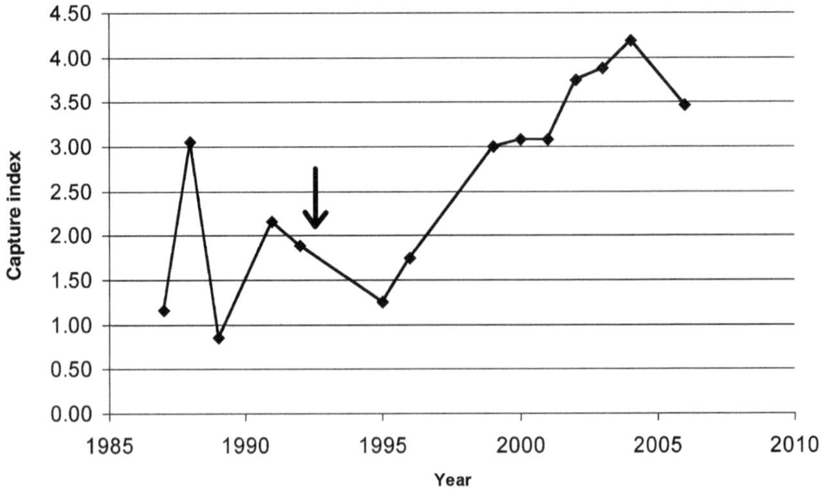

Fig. 2.1 Trapping index (number of individuals per 10 trapnights) of annual fall monitoring for the Palisades, New Jersey population of Allegheny woodrats (*Neotoma magister*). The arrow indicates the introduction of raccoon rabies and the assumed reduction of the raccoon population in 1992 (updated from LoGiudice 2006)

Historical–Ecological Hypothesis

The three hypotheses described in the preceding sections might be more coherent if taken in an historical–ecological context. It is often easy to forget how much the wildlife community of the 19th and early 20th centuries differed from that which is present today (Foster 2002, Foster et al. 2002). Most large mammals and birds in the northeastern North America had been hunted to near extirpation by 1800 and did not recover (if at all) until the mid-20th century (Rhoads 1903, Spurr and Barnes 1980, Eaton 1992, McCabe and McCabe 1997, Foster et al. 2002). Thus, during early periods of reduced food resources, the woodrat would have been experiencing relaxed competition from large, highly mobile mast consumers such as white-tailed deer, black bears, and wild turkeys. Subsequently, the recovery of these species would have reduced the share of high quality foods available to woodrats (Eaton 1992, McShea and Schwede 1992).

One can only surmise the impacts of changes in the predator community since the main enemies of woodrats and the influence of predation on woodrat demographics are unknown. Nevertheless, both the elimination of common predators (Foster et al. 2002) and the recovery of many of these species are likely to have had major influences on woodrat populations. The recovery of smaller predators such as mustelids, raccoons, and foxes in the absence of larger predators may mean that woodrats are now subjected to higher predation pressures than any time in their history.

Hypothetical Chronology

It seems unlikely that any one factor could singly be causing the range-wide reductions of the Allegheny woodrat. Rather than being mutually exclusive, the effects of each of these factors may have been working in some sites alone, in some additively, and in some synergistically. It is possible that a concatenation of these changes may have been impacting woodrat populations from the mid-1800s or before, ultimately causing the population collapses that drew human attention in the 1970s and beyond.

The following is a *hypothetical* timeline (Fig. 2.2) of the decline of the Allegheny woodrat in its northeastern range (NY, NJ, CT) where it is practically extirpated. The events in this timeline are supported by the evidence just discussed and by conjecture. The timeline is based on the dates of the historical–ecological fluctuations that might have affected woodrats and on observations of naturalists and scientists. The chronology suggests that the slope of this decline has not been constant, but rather that woodrat populations have deteriorated in a stepwise fashion, culminating in local extinctions that may have been perceived as sudden to human observers.

The intense deforestation during the 18th and 19th centuries is likely to have had a negative influence on woodrats by removing trees that provided food resources and blocking dispersal routes. This is consistent with the observations of Kalm in 1759 (cited in Rhoads 1894). However, concurrent elimination of predators and competitors may have compensated for such losses. By the time of the maximum forest clearance in the mid-1800s (Foster et al. 2002), large predators and competitors

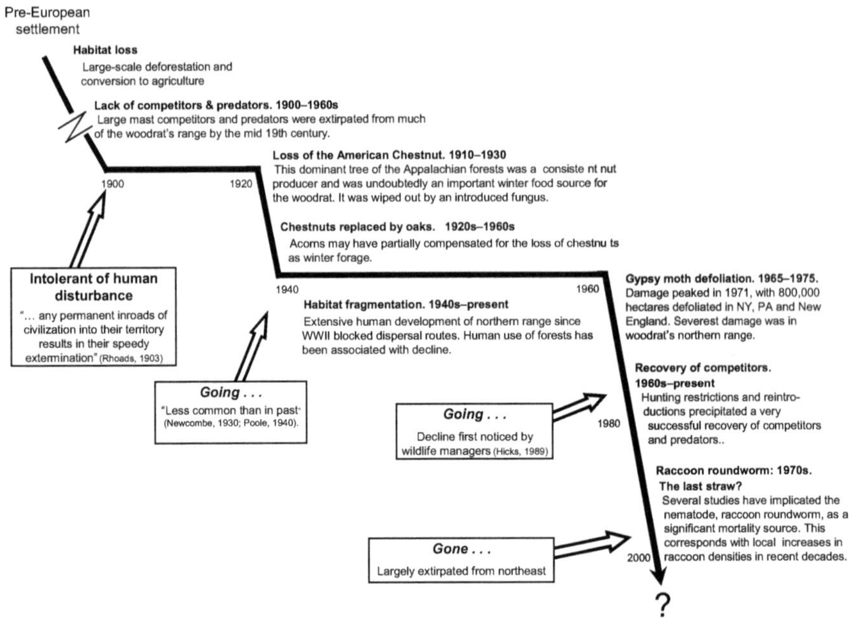

Fig. 2.2 Hypothetical chronology and population trajectory of Allegheny woodrat (*Neotoma magister*) populations in northeastern North America, from a pre-European settlement high in the upper left to the current low in the lower right. The impacts of events deleterious to woodrat survival may have been temporarily offset by simultaneous events that enhanced woodrat survival probabilities, causing a temporal disconnect between the onset of deleterious events and the population response. Actual effects of these events on population size and the precise timing of the decline are unknown and are not implied by the graphics (adapted from LoGiudice 2006)

were functionally absent from the Allegheny woodrat's habitat possibly allowing its populations to stabilize.

The destruction of the American chestnut with its reliable annual chestnut crop almost certainly impacted the woodrat and set another decline in motion. Woodrats appear to respond to food shortages by reduced reproductive output which might have increased the frequency of local extinctions. By 1920, most American chestnuts were eliminated from Connecticut, New York, and Pennsylvania and by 1930 in Ohio, Maryland, and Indiana (Campbell and Madden 1990). In many places, understory oaks filled in the gaps left by dying chestnuts (Woods and Shanks 1959, Day and Monk 1974) allowing woodrats to exploit periodic acorn crops and sustain their numbers in the absence of large mast competitors. Undoubtedly, woodrats were lost from marginal habitats and areas in which nonmasting species filled in the canopy, serving to isolate remaining populations even more. This timing is consistent with observed population declines (Newcombe 1930, Poole 1940) and may have initiated another period of stability, possibly at lower population levels. However, this stability did not last long as the enactment of strict hunting restrictions and deer and turkey reintroductions caused a rapid recovery of these two mast consumers. In the United States, populations of wild turkeys saw dramatic increases, from about 130,000 in the 1940s to 1.25 million in 1970 to 4 million in 1990 (Eaton 1992). In many places, the recovery of deer populations was even more dramatic, going from complete extirpation to dramatic overpopulation within 50 years. The recovery of the deer herd in New York/New Jersey was well underway by the 1960s (McCabe and McCabe 1997). Current deer densities frequently exceed 10 deer/km^2 and have been documented at four times that in local areas. Grazing by deer at these densities can prevent oak regeneration (Healy 1997) and reduce small mammal populations (McShea and Schwede 1992). At this time black bear populations were also beginning to recover and direct competition with these three species likely reduced the woodrat's share of the mast crop. Finally, mortality from recovering populations of great horned (*B. virginianus*) and barred (*Strix varia*) owls and mammalian mesopredators (Foster et al. 2002) could have had a large impact on woodrat demographics (Balcom and Yahner 1996).

At the same time that competition and predation were increasing, the woodrat range was being fragmented by human development. This likely blocked dispersal routes and prevented recolonization of extirpated sites, interrupting the metapopulation dynamics. The early disappearance of metapopulations in the lower Hudson River Valley, one of the most densely developed regions in the woodrat range, implies the importance of fragmentation. A similar decline is evident in the central Appalachians north of the Potomac River, also an area of long-term human impact (Ford et al. 2006).

As turkey and white-tailed deer populations surged in the 1970s, another competitor for the oak mast was developing. After a slow start from its introduction near Boston in the late 1860s, the gypsy moth expansion had picked up and was rapidly consuming forests through the northeast. In New York, Pennsylvania, and New England, damage peaked in 1970 with 800,000 ha defoliated. Extremely high levels of oak mortality were reported in New York, Pennsylvania, Connecticut, and New Jersey (McManus and McIntyre 1981). This is consistent with the first discovery of the woodrat decline.

By the 1960s, raccoon populations appear to have reached very high densities in the northern portion of the woodrat's range. Although quantitative data are scarce, this consummate human-adapted species has benefited from the transition from pastoral to suburban land uses and the resultant reduction in hunting that followed World War II (Riley et al. 1998). Prominent New York naturalist Dan Smiley documented a dramatic increase in raccoons, commenting that they were rare in the 1920s and a nuisance by the 1970s (Smiley 1977). Changes in raccoon habitat use might also have caused more contact between the two species than had been the case during much of their evolutionary history. The loss of large den trees may have pushed raccoons to seek shelter in the talus. The high degree of resistance to *B. procyonis* found in the white-footed mouse may imply an evolved response to a long history of exposure (Sheppard and Kazacos 1997). This seems likely since raccoons and white-footed mice typically share woodland habitats. Conversely, lack of evolved resistance to this parasite in the Allegheny woodrat may indicate that, historically, exposure was less common (Wakelin 1994). To end our timeline, by the late 1970s woodrat populations appear to have been beleaguered by fragmentation and low recruitment making them highly vulnerable to stochastic mortality sources. Raccoon roundworm is likely to have played this role in New York and New Jersey at least, serving as the final straw and causing the rapid collapses observed in these states.

The forces described in the proposed chronology (Fig. 2.2) need not have acted independently on woodrat populations. Food shortage and increased interspecific competition may have caused death directly by starvation or indirectly, by forcing animals to forage farther for food, exposing them to predation and/or the elements. They may have acted synergistically with parasites by increasing foraging in raccoon latrines and causing chains of mortality as one individual after another occupied a contaminated den site. Increasing intraspecific competition for scarce mast might have caused more agonistic interactions among woodrats. Loss of habitat or dispersal routes could add to these woes. It appears that dispersal in this species is so limited (Wood Chapter 3) that the loss of a single centrally located and productive subpopulation could precipitate the collapse of an entire metapopulation. There may also be undetected threshold effects operating, where recolonization is virtually impossible if metapopulation occupancy drops below a certain limit. More information on the dispersal abilities of woodrats would help to determine the proper spatial scale in which to consider a metapopulation and would facilitate the modeling of metapopulation dynamics.

The small size of individual woodrat subpopulations suggests that they are quite vulnerable to extirpation. Thus, the species has a precarious population structure, with dispersal, mortality, and recruitment delicately balanced to maintain metapopulations. In the past, stability of the metapopulation structure has allowed long-term survival even in the face of stochastic subpopulation extinctions. However, any factor that leads to an increase in mortality or a reduction in dispersal or recruitment could exceed a viability threshold and result in metapopulation collapse. The hypotheses presented here potentially increase mortality (food shortage, increased predation, *B. procyonis*), reduce fecundity (irregular food availability, increased competition), and reduce recolonization frequency (habitat fragmentation and loss of subpopulations), so any combination of these factors could destabilize metapopulations and cause local extinction.

Conservation and Management of Declining Species

The Allegheny woodrat is emblematic of many endangered and threatened species. Humans have been changing habitats in complex ways and on multiple levels for centuries. In the process, we set off ecological chain reactions that may be invisible to us until suddenly, and perhaps too late, they attract our attention. Too often, these intricate relationships are all but irreparable without huge commitments of human resources. However, it is useful to examine such cases in the hope that we can learn to recognize the warning signs and avoid making the same mistakes in the future.

What lessons can be learned from studying the decline of the Allegheny woodrat? First, that it is instructive to thoroughly consider multiple hypotheses for a species' decline before discarding any, regardless of how insignificant they may seem. It is not safe to assume that a single factor is responsible for a decline or that the impact of a given factor is too small to be relevant. In a thorough review of extinctions, Pimm (1996) concluded that most extinctions were caused by synergistic interactions between multiple factors rather than being attributable to a single cause. Synergistic, additive, or compensatory effects may exaggerate the impact of a seemingly mild insult or mask the importance of a major factor, so ignoring a minor factor may be fatal. Threshold effects may also be operating, such that incremental changes have little or no impact until the threshold is exceeded and a large decline ensues. Of course, it is not practical or even possible to thoroughly test and remediate every potential factor. If the collapse of the woodrat range is due to a multitude of factors, how does one identify the "last straw"? This is enormously challenging and requires coordination and collaboration. No one agency, lab, or investigator can adequately explore all possible hypotheses. Thus, it would be helpful to form a working group of state and federal agencies, academics, and other interested parties. This group should work cooperatively to produce and disseminate research "punch lists" of factors that should be investigated in as many places and with as many methods as money and interest allow. Often, small pilot projects can help to identify promising hypotheses. When possible, multiple factors should be investigated simultaneously to test for interactions. And of course, members of the group should share results as promptly as practicable.

Next, we should study the anomalous populations. It is often as instructive to investigate the successes as the failures. Just as we can learn by asking the question: "Why is this population declining?" We can also learn by asking: "Why is this population persisting?" Isolated populations of Allegheny woodrat remain throughout the range, surviving against all the odds as other nearby populations have disappeared. While they exist, they present opportunities to learn about the key factors for population persistence and to discover possible interventions. Comparative studies between these sites would be most instructive. Of course, we should remember the precarious position of these populations which makes manipulative studies quite dangerous, as a small perturbation might be all that is necessary to drive such a population to extinction. Especially with habitat specialists and other vulnerable species, even small population declines may be followed by massive collapse. Close monitoring of representative populations is necessary and a swift reaction should follow if a decline is detected. This means developing an emergency action plan using the best data available.

Finally, it is useful to take the historical–ecological approach and be aware of historical patterns in resource availability, predation, and competition to better understand the varying forces to which endangered species have been subjected. We should keep history in mind and beware of eliminating a potential factor because it does not currently seem important. We must conduct the thought experiment of going back 50, 100 and 200 years to determine as best we can what the species was experiencing at those times. By recognizing the importance of historical patterns and considering all plausible hypotheses, we can help to expand our understanding of what imperiled species have been through and what they may need to persist.

References

Anagnostakis, S. L. 2000. Hemming's Chinese chestnuts. Annual Report of the Northern Nutgrowers Association 90:32–35.

Balcom, B. J., and Yahner, R. H. 1996. Microhabitat and landscape characteristics associated with the threatened Allegheny woodrat. Conservation Biology 10:515–525.

Birch, G. L., Feldhamer, G. A., and Dyer, W. G. 1994. Helminths of the gastrointestinal tract of raccoons in southern Illinois with management implications of *Baylisascaris procyonis* occurrence. Transactions of the Illinois State Academy of Science 87:165–170.

Brewer, L. G. 1995. Ecology of survival and recovery from blight in American chestnut trees (*Castanea dentata*) in Michigan. Bulletin of the Torrey Botanical Club 122:40–57.

Campbell, C. L., and Madden, L. V. 1990. Introduction to Plant Disease Epidemiology. John Wiley and Sons, New York.

Castleberry, N. L., Castleberry, S. B., Ford, W. M., Wood, P. B., and Mengak, M. T. 2002. Allegheny woodrat (*Neotoma magister*) food habits in the central Appalachians. American Midland Naturalist 147:80–92.

Chamblin, H. D., Wood, P. B., and Edwards, J. W. 2004. Allegheny woodrat (*Neotoma magister*) use of rock drainage channels on reclaimed mines in southern West Virginia. American Midland Naturalist 151:346–354.

Cudmore, W. W. 1983. The distribution and ecology of the eastern woodrat, *Neotoma floridana*, in Indiana. Ph.D. Dissertation, Indiana State University, Terre Haute, Indiana.

Day, F. P., Jr., and Monk, C. D. 1974. Vegetation patterns on a southern Appalachian watershed. Ecology 55:1064–1074.

Diamond, J. 1989. Overview of recent extinctions. *In* Conservation for the Twenty-First Century, eds. D. Western and M. C. Pearl, pp. 37–41. Wildlife Conservation International, New York.

Eaton, S. W. 1992. Wild turkey (*Meleagris gallopavo*). *In* The Birds of North America, Number 22, eds. A. Poole, P. Stettenheim and F. Gill, pp. 1–28. Philadelphia Academy of Natural Sciences, Philadelphia, Pennsylvania.

Eberhard, M. L., Nace, E. K., Won, K. Y., Punkosdy, G. A., Bishop, H. S., and Johnston, S. P. 2003. *Baylisascaris procyonis* in the metropolitan Atlanta area. Emerging Infectious Diseases 9:1636–1637.

Eberhardt, R. W., Foster, D. R., Motzkin, G., and Hall, B. 2003. Conservation of changing landscapes: vegetation and land-use history of Cape Cod National Seashore. Ecological Applications 13:68–84.

Elias, T. S. 1980. The Complete Trees of North America: Field Guide and Natural History. Van Nostrand Reinhold, New York.

Ford, W. M., Castleberry, S. B., Mengak, M. T., Rodrigue, J. L., Feller, D. J., and Russell, K. R. 2006. Persistence of Allegheny woodrats *Neotoma magister* across the mid-Atlantic Appalachian Highlands landscape, USA. Ecography 29:745–754.

Foster, D. R. 2000. From bobolinks to bears: interjecting geographical history into ecological studies, environmental interpretation, and conservation planning. Journal of Biogeography 27:27–30.

Foster, D. R. 2002. Thoreau's country and a historical-ecological approach to conservation of the New England landscape. Journal of Biogeography 29:1537–1555.

Foster, D. R., Motzkin, G., Bernardos, D., and Cardoza, J. 2002. Wildlife dynamics in the changing New England landscape. Journal of Biogeography 29:1337–1357.

Gompper, M. E., and Hackett, H. M. 2005. The long-term, range-wide decline of a once common carnivore: the eastern spotted skunk (*Spilogale putorius*). Animal Conservation 8:195–201.

Hall, J. P. 1987. Survey of the woodrat in eastern Pennsylvania. Contract 878903, Final report to the Pennsylvania Game Commission, Harrisburg, Pennsylvania.

Hassinger, J. D., and Butchkoski, C. 1999. Eastern woodrat research/management. Job Code 71801, Annual report to the Pennsylvania Game Commission, Harrisburg, Pennsylvania.

Healy, W. M. 1997. Influence of deer on the structure and composition of oak forests in central Massachusetts. *In* The Science of Overabundance: Deer Ecology and Population Management, eds. W. J. McShea, H. B. Underwood and J. H. Rappole, pp. 249–266. Smithsonian Institution Press, Washington, D.C.

Healy, W. M., Gottschalk, K. W., Long, R. P., and Wargo, P. M. 1997. Changes in eastern forests: chestnut is gone, are the oaks far behind? Transactions of the North American Wildlife and Natural Resources Conference 62:249–263.

Hicks, A. P. 1989. Whatever happened to the Allegheny woodrat? The New York State Conservationist 43:34–38.

Hudson, C. M., Johnson, S. A., Holbrook, A. L., Walker, H. D., and Walker, R. M. 2006. 2005 Allegheny woodrat monitoring program. Indiana Department of Natural Resources, Division of Fish and Wildlife, Bloomington, Indiana.

Jacobson, J. E., Kazacos, K. R., and Montague, F. H., Jr. 1982. Prevalence of eggs of *Baylisascaris procyonis* in raccoon scats from an urban and rural community. Journal of Wildlife Diseases 18:461–464.

Johnson, S. A. 2002. Reassessment of the Allegheny woodrat (*Neotoma magister*) in Indiana. Proceedings of the Indiana Academy of Science 1:56–66.

Jones, C. G., Ostfeld, R. S., Richard, M. P., Schauber, E. J., and Wolff, J. O. 1998. Chain reactions linking acorns to gypsy moth outbreaks and Lyme disease risk. Science 279: 1023–1026.

Justice, F., and Smith, F. A. 1992. A model of dietary utilization by small mammalian herbivores, with empirical results for *Neotoma*. American Naturalist 139:398–416.

Kazacos, K. R. 2001. *Baylisascaris procyonis* and related species. *In* Parasitic Diseases of Wild Mammals, Second edition, eds. W. M. Samuel, M. J. Pybus and A. A. Kocan, pp. 301–341. Iowa State University Press, Ames, Iowa.

Liebhold, A. M., Elmes, G. A., Halverson, J. A., and Quimby, J. 1994. Landscape characterization of forest susceptibility to gypsy moth defoliation. Forest Science 40:18–29.

LoGiudice, K. 2001. Latrine foraging strategies of two small mammals: implications for the transmission of *Baylisascaris procyonis*. American Midland Naturalist 146:369–378.

LoGiudice, K. 2003. Trophically transmitted parasites and the conservation of small populations: raccoon roundworm and the imperiled Allegheny woodrat. Conservation Biology 17: 258–266.

LoGiudice, K. 2006. Toward a synthetic view of extinction: a history lesson from a North American rodent. BioScience 56:687–693.

Long, D. B., Campbell, T. A., and Henke, S. E. 2006. *Baylisascaris procyonis* (Nematoda: Ascaridoidea) in raccoons (*Procyon lotor*) from Duval County, Texas. Texas Journal of Science 58:281–285.

McCabe, T. R., and McCabe, R. E. 1997. Recounting whitetails past. *In* The Science of Overabundance: Deer Ecology and Population Management, eds. W. J. McShea, H. B. Underwood and J. H. Rappole, pp. 11–26. Smithsonian Institution Press, Washington, D.C.

McClure, P. A. 1987. The energetics of reproduction and life histories of cricetine rodents (*Neotoma floridana* and *Sigmodon hispidus*). Symposium of the Zoological Society of London 57:241–258.

McClure, P. A., and Randolph, J. C. 1980. Relative allocation of energy to growth and development of homeothermy in the eastern woodrat (*Neotoma floridana*) and hispid cotton rat (*Sigmodon hispidus*). Ecological Monographs 50:199–219.

McDonald, R. A., and Harris, S. 1999. The use of trapping records to monitor populations of stoats *Mustela erminea* and weasels *M. nivalis*: the importance of trapping effort. Journal of Applied Ecology 36:679–688.

McGowan, E. M. 1993. Experimental release and fate study of the Allegheny woodrat (*Neotoma magister*). Federal aid project report W-166-E-1, New York Department of Environmental Conservation, Delmar, New York.
McManus, M. L., and McIntyre, T. 1981. Introduction. *In* The Gypsy Moth: Research Toward Integrated Pest Management, eds. C. C. Doane and M. L. McManus, pp. 1–32. U.S.D.A. Forest Service Science and Education Agency, Animal and Plant Health Inspection Service Technical Bulletin 1584, Washington, D.C.
McShea, W., and Schwede, G. 1992. Variable acorn crops: responses of white-tailed deer and other mast consumers. Journal of Mammalogy 74:999–1006.
Mengak, M. T. 2002. Reproduction, juvenile growth and recapture rates of Allegheny woodrats (*Neotoma magister*) in Virginia. American Midland Naturalist 148:145–152.
Murphy, M. F. 1952. Ecology and helminths of the Osage woodrat, *Neotoma floridana osagensis*, including the description of *Longistriata neotoma*. American Midland Naturalist 48: 204–217.
Newcombe, C. L. 1930. An Ecological study of the Allegheny cliff rat (*Neotoma pennsylvanica* Stone). Journal of Mammalogy 11:204–211.
Owen, S. F., Edwards, J. W., Ford, W. M., Crum, J. M., and Wood, P. B. 2004. Raccoon roundworm in raccoons in Central West Virginia. Northeastern Naturalist 11:137–142.
Page, L. K., Swihart, R. K., and Kazacos, K. R. 1999. Implications of raccoon latrines in the epizootiology of Baylisascariasis. Journal of Wildlife Diseases 35:474–480.
Page, L. K., Gehrt, S. D., Titcombe, K. K., and Robinson, N. P. 2005. Measuring prevalence of raccoon roundworm (*Baylisascaris procyonis*): a comparison of common techniques. Wildlife Society Bulletin 33:1406–1412.
Pennsylvania Chestnut Blight Conference. 1912. Stenographic report of proceedings, C. E. Aughinbaugh. Harrisburg, Pennsylvania.
Pennsylvania Game Commission. 1995. Allegheny woodrat site survey code manual. Pennsylvania Game Commission, Harrisburg, Pennsylvania.
Pimm, S. L. 1996. Lessons from a kill. Biodiversity and Conservation 5:1059–1067.
Poole, E. L. 1940. A life history sketch of the Allegheny woodrat. Journal of Mammalogy 21: 249–318.
Prange, S., and Gehrt, S. D. 2004. Changes in mesopredator-community structure in response to urbanization. Canadian Journal of Zoology 82:1804–1817.
Rhoads, S. N. 1894. A contribution to the life history of the Allegheny cave rat, *Neotoma magister*, Baird. Proceedings of the Academy of Natural Sciences of Philadelphia 46:213–221.
Rhoads, S. N. 1903. The Mammals of Pennsylvania and New Jersey. Privately published, Philadelphia, Pennsylvania.
Riley, S. P. D., Hadidian, J., and Manski, D. A. 1998. Population density, survival, and rabies in raccoons in an urban national park. Canadian Journal of Zoology 76:1153–1164.
Sheppard, C. H. and Kazacos, K. R. 1997. Susceptibility of *Peromyscus leucopus* and *Mus musculus* to infection with *Baylisascaris procyonis*. Journal of Parasitology 83:1104–1111.
Shotola, S. J., Weaver, G. T., Robertson, P. A., and Ashby, W. C. 1992. Sugar maple invasion of an old-growth oak-hickory forest in Southwestern Illinois. American Midland Naturalist 127:125–138.
Sikes, R. S. 1995. Maternal response to resource limitation in eastern woodrats. Animal Behaviour 49:1551–1558.
Smiley, D. 1977. Raccoon. Research report to the Mohonk Trust, New Paltz, New York.
Spurr, S. H., and Barnes, B. V. 1980. Forest Ecology. Third edition. John Wiley and Sons, New York.
Wakelin, D. 1994. Host populations: genetics and immunity. *In* Parasitic and Infectious Diseases, Epidemiology and Ecology, eds. M. E. Scott and G. Smith, pp. 83–100. Academic Press, New York.
Whitaker, J. O., Jr., and Hamilton, W. J., Jr. 1998. Mammals of the Eastern United States. Cornell University Press, Ithaca, New York.

Wolff, J. O. 1996. Population fluctuations of mast-eating rodents are correlated with production of acorns. Journal of Mammalogy 77:850–856.

Woods, F. W., and Shanks, R. E. 1959. Natural replacement of chestnut by other species. Ecology 40:349–361.

Wright, J., and Kirkland, G. L. 2000. A possible role for chestnut blight in the decline of the Allegheny woodrat. Journal of the American Chestnut Foundation 8:30–35.

Zielinski, W. J., Slauson, K. M., Carroll, C. R., Kent, C. J., and Kudrna, D. G. 2001. Status of American martens in coastal forests of the Pacific states. Journal of Mammalogy 82:478–490.

Section II
Ecology

The conservation of a declining species is, by necessity, a form of crisis management. In the case of the Allegheny woodrat (*Neotoma magister*), the ability of scientists to address the crisis was initially hindered by lack of basic information concerning the ecology of the species. Chapters in this section summarize our current knowledge of the ecology of the Allegheny woodrat—virtually all of which has been accumulated after it was recognized as a declining species. Careful examination of the findings summarized in these chapters not only provides insight into factors that might have contributed to the decline of *N. magister*, but also highlights aspects of the species' ecology that must be considered in efforts to promote the recovery of this species.

Birth, death, and migration rates are basic processes in any population, but understanding of these is complicated when a species' distribution is discontinuous and local populations are relatively isolated. In Chapter 3, Petra Wood summarizes population-level studies of Allegheny woodrats conducted by her research group in West Virginia. She combines data on demographics, dispersal, and genetics to provide convincing evidence that woodrats exhibit a metapopulation structure. These metapopulations consist of individual rock outcrops representing subpopulations (i.e., local populations) with limited dispersal among outcrops. Although most demographic data have been derived from subpopulations, the inferences from these must ultimately apply at the metapopulation level.

In Chapter 4, Steven Castleberry provides an overview of the factors influencing habitat selection by Allegheny woodrats. He also summarizes studies reporting home ranges and foraging movements of the Allegheny woodrat. The results of these studies suggest that the habitat specificity of *N. magister* necessitates greater foraging distances compared to other species. Results of these studies also suggest a potential interaction among seasonal energetic needs, food quality, and human disturbance. Thus, movement patterns of individuals reflect events that influence survival and reproduction within a local population.

Any process that limits the number of woodrats occupying individual rock outcrops also will influence the dynamics of populations at a larger scale. Although behavior may influence population dynamics, it is only recently that this factor has received consideration in the context of conservation. In Chapter 5, John Peles and Janet Wright report the results of long- and short-term studies of behaviors related to den use and den selection. Their findings suggest that behavioral characteristics need to be considered as potentially limiting factors for local population size. They present a model for the potential effects of woodrat behaviors on the size of local populations as well as metapopulation processes.

Food quality and quantity also may influence population dynamics, survivorship, and reproduction within local populations. In Chapter 6, Nikole and Steven Castleberry summarize the known components in the diet of woodrats and the contents of food caches. As might be expected, hard mast is an important dietary component during certain seasons. The authors discuss the implications of temporal variation in the quantity and quality of mast crops as a factor contributing to the historical decline of the woodrat as well as the dynamics of extant populations. At the same time, they point out that much remains to be learned about the importance of resource diversity in woodrat diets.

John D. Peles
Janet Wright

Chapter 3
Woodrat Population Dynamics and Movement Patterns

Petra Bohall Wood

Introduction

The Allegheny woodrat (*Neotoma magister*) is a habitat specialist that typically inhabits discrete rock patches embedded in a forest dominated matrix. Rocky habitat provides nest sites and protection from predators (Newcombe 1930, Poole 1940). High habitat specificity combined with a patchy distribution of rock habitats across the landscape results in a distributional pattern of small and presumably somewhat isolated clusters of individuals representing subpopulations within larger metapopulations.

Several lines of evidence are consistent with the metapopulation model for Allegheny woodrats. Metapopulation theory suggests that two key factors are necessary for metapopulations to occur: a spatially discrete distribution and a nontrivial probability that extinction could occur in one or more local patches (McCullough 1996). Within a metapopulation, subpopulations are linked by dispersal, and habitat patches are not so isolated that recolonization is prevented (Wiens 1996). With a metapopulation structure, we would not expect all habitat patches to be occupied at all times. For example, between September 1995 and October 1996, Myers (1997) captured Allegheny woodrats at 25 of 69 rock outcrops trapped. Although trapping sites were distributed throughout his north-central West Virginia study area, captures had a clumped distribution. Smaller and more isolated rock sites tended to have fewer captures. However, most females that inhabited even isolated rock outcrops showed signs of lactation, so males apparently were able to locate and breed with them throughout the season (Myers 1997).

Within metapopulations, subpopulations have a nontrivial probability of extinction (McCullough 1996). Thus, we might expect yearly variability in capture rates at individual rock sites as subpopulations move toward or away from extinction. Annual trapping during 1995–2003 revealed annual differences among outcrops in

P.B. Wood
U.S. Geological Survey, West Virginia Cooperative Fish and Wildlife Research Unit, Division of Forestry and Natural Resources, West Virginia University, Morgantown, WV 26506

J.D. Peles, J. Wright (eds.), *The Allegheny Woodrat*, DOI: 10.1007/978-0-387-36051-5_3,

capture rates (number of individuals per trap night) of all woodrats, adult females, and juveniles (Manjerovic 2004). The small size of typical woodrat subpopulations (1–5 individuals per rock site) makes them susceptible to stochastic fluctuations which could lead to local extinction. Capture rates of adult females significantly decreased at 3 of 12 outcrops across years, although juvenile capture rates increased at two outcrops (Manjerovic 2004).

Metapopulation theory implies that woodrats might temporarily disappear but then recolonize sites. Manjerovic (2004) found significant genetic relatedness among individuals from certain individual outcrops that occurred in a linear pattern. This suggests that the connectivity between specific outcrops distributed in linear patterns facilitates movements and dispersal which enhances recolonization of outcrops. A metapopulation structure has further been confirmed by other genetic studies (Castleberry et al. 2002b) and by radiotelemetry and mark-recapture assessment of dispersal and movements (Myers 1997, J. Zuck and P. Wood unpublished data).

In this chapter, I review and summarize literature on movement and dispersal patterns, demography, and population dynamics of Allegheny woodrats. I relate this information to metapopulation structure and draw implications for conservation of this and other declining species.

Movements and Dispersal

Allegheny woodrats are capable of traveling relatively long distances regardless of sex and age. In West Virginia, Castleberry et al. (2001a) radio-tracked an individual that moved >1000 m in 1 h, while Myers (1997) recaptured a juvenile female that traveled 3.3 km from her natal site in 1.5 months. An adult male that had been radio-tracked for approximately 24 months in Pennsylvania suddenly moved 6 km along a ridge top to a new site (J. Wright unpublished data). A dead radiocollared woodrat was recovered 3.2 km from its release site in New York (Beans 1992). Although Allegheny woodrats are capable of long-distance movements, the incidence appears low and occurs most often in adult males. Of 83 males captured in 1999–2000, 9% ($n = 8$ individuals) made long-distance movements of >200 m as adults; recapture distances within and across years were 270–1300 m, while only 1% of 84 adult females captured ($n = 1$) made long-distance movements (J. Zuck and P. Wood unpublished data).

Mammalian movements are influenced by age, breeding season, and habitat. Both sexes of mammalian juveniles make exploratory movements that allow the juvenile to become familiar with surrounding areas and possibly to discover vacant areas of suitable resources (McShea and Madison 1992); presumably this applies to Allegheny woodrats. Adult Allegheny woodrats can travel considerable distances beyond the protection of rock sites (Wright 1996) in search of food and mates (Poole 1940) as do dusky-footed woodrats (*N. fuscipes*; Linsdale and Tevis 1951) and bushy-tailed woodrats (*N. cinerea*; Topping and Millar 1996). Cranford (1977) observed male *N. fuscipes* to increase their home ranges during the female estrous cycle. Male woodrats of several species have large home ranges (Goertz 1970,

Topping and Millar 1996, Conditt and Ribble 1997, Castleberry et al. 2001a) that may overlap with numerous female home ranges (Lynch et al. 1994). Males also are more likely to move between rock sites within and among years, presumably in search of breeding females. During the breeding season, a radiocollared male Allegheny woodrat moved between its den and a rock site inhabited by females approximately 600 m away (Myers 1997). A radiocollared male moved to an adjacent rock site over 500 m away and returned each time to its den site within 24 hours on three different occasions, while another male was captured at four different adjacent rock sites during 1999–2000 (J. Zuck and P. Wood unpublished data).

In patchy landscapes, direction and distance of mammalian movements may reflect the size, spacing, and quality of habitat (Topping and Millar 1996, Diffendorfer et al. 1999). Movements are seldom random, but tend to follow pathways, drainage systems, and topographic features; Allegheny woodrat recaptures on our study area in West Virginia often occurred along rock outcrops arranged in a linear pattern. Areas of connectivity are particularly important to metapopulations where fragmented habitats and other anthropogenic activity have created habitat isolation. Movement of woodrats between rock patches decreases the likelihood that a metapopulation would go extinct, allowing for gene flow and recolonization after local extinction.

Populations isolated by barriers normally cannot achieve long-term viability and eventually go extinct. Highways, urbanization, and agricultural lands appear to be barriers to Allegheny woodrat movements (LoGiudice 2006), but rivers do not. Based on genetic data, populations dissected by two rivers approximately 100 m in width showed little evidence of geographic isolation (Castleberry et al. 2002b) suggesting that mid-sized rivers are not a complete barrier to dispersal. Manjerovic (2004) found greatest numbers of effective migrants between subpopulations that were close together but separated by the Cheat River in West Virginia, suggesting that the river does not impede migration between populations.

Some long-distance movements qualify as dispersal movements, which are important for maintaining metapopulation structure. Dispersal is a behavioral process that describes patterns of movements by individuals between and among populations. A widely used definition of dispersal by Howard (1960) states that "dispersal of an individual vertebrate is the movement the animal makes from its point of origin to the place where it reproduces or would have reproduced had it survived and found a mate." Dispersal is considered characteristic of juveniles that typically do not return to the natal area after departure (termed *natal dispersal* after Greenwood 1980), although dispersal by adults among breeding areas can also occur (*breeding dispersal* after Greenwood 1980). In contrast, philopatry occurs when individuals settle in or near their natal area.

Dispersal affects the distribution and abundance of organisms and the structure of populations. Consequences include the spacing of individuals, exchange of genetic material among populations, local patterns of relatedness and age structure, survival and recruitment, population and species range expansions, and social behavior (Greenwood 1980). Dispersal also is a key component in metapopulation dynamics linking subpopulations in spatially discrete habitat patches (McCullough 1996). In small mammal populations, dispersal has a significant regulatory role (Lidicker 1975) in the maintenance of genetic variability and in gene flow among subpopulations (Howard 1960). Until recently, little was known about Allegheny woodrat dispersal, particularly juvenile dispersal. However, it has been assumed that, like other mammalian species (Greenwood 1980), male woodrats disperse more often and at greater distances than females.

Despite the obvious importance of dispersal to maintenance of Allegheny woodrat metapopulations, relatively little was known about which sex, age, and seasons are most likely to be involved. We collected movement and dispersal data in a mark-recapture study in north-central West Virginia during 1995–2004 at a variable number of sites trapped 1–3 times each year (Myers 1997, Manjerovic 2004, J. Zuck and P. Wood unpublished data) and in a radiotelemetry study during 1999–2000 (J. Zuck and P. Wood unpublished data). In 1999–2000, woodrats were trapped for two consecutive nights 1–2 times per year at 19 rock sites from May to September; 21 of the juveniles captured were fitted with radiocollars. Woodrats were radio-tracked to their diurnal dens 2–4 times per week between June 1999–January 2000 and May–December 2000. Also, these individuals were radio-tracked approximately one night per week to ensure that they were still alive by listening to a fluctuating signal indicating that the animal was moving.

Dispersal movements in juvenile Allegheny woodrats appear to be uncommon, a result suggesting low incidence of natal dispersal. We detected no long-distance movements in juveniles during 1999–2000 based on 18 radiocollared juveniles or via mark-recapture records of 81 juveniles (38 M, 43 F). Recaptures of these juveniles always occurred at the rock site where originally captured and none of these individuals were confirmed to have dispersed as juveniles. Attempts to recapture dispersers at rock sites outside the study areas in 2000 failed to document dispersal movements from the natal rock site. Between 1995 and 2004, only one instance of juvenile dispersal was documented for 169 individual juveniles captured; this was a juvenile female recaptured by Myers (1997) that moved 3.3 km from her natal site.

Juvenile philopatry also was evident when examining geographic distance between female Allegheny woodrats and their offspring. Manjerovic (2004) identified maternal-offspring pairs through genetic analyses and then measured geographic distances between females and offspring in each pair based on time of capture. These distances were measured for each male and female juvenile captured and in subsequent years for each juvenile recaptured as an adult. Low average distance and low variation in distance from maternal woodrats for juvenile females and males (Fig. 3.1) indicated that both sexes of juveniles exhibit philopatry during their juvenile year. Once juveniles become adults, distance from maternal parent is much greater for male than female offspring. The range of distances between maternal-offspring pairs indicates that both sexes are potential dispersers although female offspring are more likely to remain in close proximity to the maternal parent. The large change in geographic distance for male offspring when juveniles were recaptured as adults suggests that woodrats move away from their natal site between their juvenile and adult years.

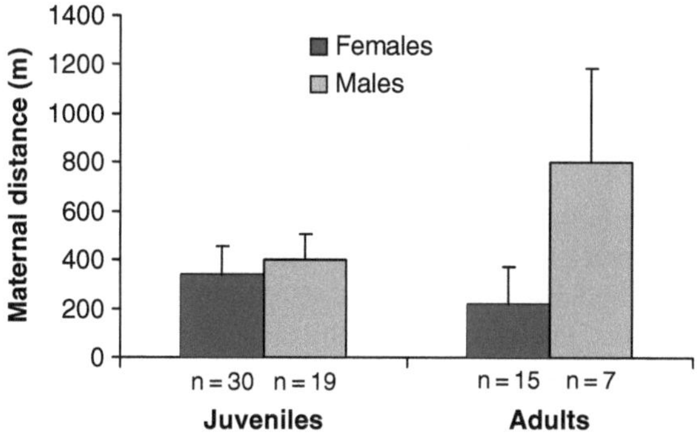

Fig. 3.1 Mean estimated distance (m) from mother to female and male offspring in Allegheny woodrats (*Neotoma magister*) captured as juveniles and those recaptured as adults. Bars represent standard error. Adapted from Manjerovic (2004)

Similarly, based on recapture data, natal dispersal appeared to occur the following spring with the timing possibly related to higher food availability in spring. We found that two first-year adult males had moved considerable distances (1600 and 2000 m) from where they were captured the previous year as juveniles. One of these first-year adult males was captured as a juvenile in 1999, was recaptured as an adult in 2000 at an adjacent rock site 300 m from its natal area, then traveled 1300 m to another rock site where it remained until it was preyed upon approximately four weeks later. Another male that was initially captured as a juvenile in 1998 was recaptured during the breeding season of the following year at a site 2000 m away but was never recaptured after being trapped at this new location. It is likely that these individuals had dispersed from their natal area.

As expected for juvenile woodrats, we found in our studies that young males appear to disperse more often than females. During 1995–2004, 14 male Allegheny woodrats originally marked as juveniles were recaptured as adults with seven recaptured at a different rock site. In contrast, three of 23 juvenile females recaptured as adults were at a different rock site. For small mammals in general, females generally are philopatric (Greenwood 1980, Boonstra et al. 1987) to their natal area.

Female woodrats also are generally philopatric as adults and exhibit no adult breeding dispersal movements. In *N. cinerea*, females tend to spend their entire reproductive life at their natal outcrop (Moses and Millar 1994). Female Allegheny woodrats appear to exhibit a similar pattern. Manjerovic (2004) detected low geographic distance between maternal woodrats and their adult female offspring (Fig. 3.1). Of 84 females captured in 1999 and 2000, $<1\%$ ($n = 1$) had movements of >200 m in our study. This female, initially marked as a juvenile in June 1999, was recaptured 13 times at the same rock site up to September 2001 although one trap station was approximately 320 m away from the initial capture site. This long-distance movement was probably a foraging movement as Topping and Millar (1996) found that female *N. cinerea* will travel up to 470 m from the nest in search of food. No movements >80 m were recorded for the other 83 adult females in our study, with many females captured at the same trap station within and across years. Other authors (Pearson 1952, Myers 1997) have reported short recapture distances for females of various woodrat species. The smaller home range size in general for female Allegheny woodrats in comparison to males (Castleberry et al. 2001a, Castleberry Chapter 4) also supports philopatric behavior and lack of long distance movements for adult females.

Woodrats in general exhibit high site fidelity and low dispersal rates (Cranford 1977, Moses and Millar 1994, Conditt and Ribble 1997, Myers 1997). Due to high turnover rates of adults, juveniles may establish themselves within their natal ranges or in adjacent home ranges. Such a pattern has been observed in Allegheny woodrats (Myers 1997) and in *N. fuscipes* (Cranford 1977). Myers (1997) suggested that higher recapture rates for juvenile females than males indicated philopatry of juvenile females who often took over their maternal den if the adult female disappeared. Similarly, juvenile females of *N. cinerea* are philopatric, often remaining at their natal outcrop and breeding close to the maternal den site (Moses and Millar 1994). Juvenile small mammals that settle within their natal area gain advantages through access to parental resources, familiarity with feeding, refuge, and den sites, and close proximity to tolerant relatives (Waser and Jones 1983, Boonstra et al. 1987); this likely applies to woodrats as well. Survivorship is expected to be greater for individuals that move relatively short distances from their place of birth. Allegheny woodrats may disperse by shifting their centers of activity within familiar areas over a period of time, a pattern which would only be detected through intensive trapping over a large area.

The low rates of dispersal in Allegheny woodrats suggested by field data are supported by genetic data. Low gene flow estimates between subpopulations in north-central West Virginia (1.0–2.6 effective migrants per generation, Castleberry et al. 2002b; 0.2–3.1, Manjerovic 2004) coupled with significant genetic differentiation among subpopulations (Castleberry et al. 2002b, Manjerovic 2004) suggest that dispersal is limited. The lowest effective migration rate was between subpopulations separated by the greatest distance (Castleberry et al. 2002b, Manjerovic 2004). Thus, geographic distance can be a meaningful barrier to dispersal. Subpopulations at individual rock outcrops are measurably differentiated but do show gene flow, so that differentiation is somewhat counteracted by dispersal; some woodrats are moving between and among outcrops. However, only limited genetic communication appears to occur among subpopulations in the form of nomadic males and the occasional disperser. During 1999–2000, only one documented movement occurred among subpopulations in our study area, a male that traveled 2 km, whereas all other movements detected were within a genetic subpopulation as defined by Castleberry et al. (2002b). Additionally, juveniles recaptured as adults generally were <200 m from their natal area (J. Zuck and P. Wood unpublished data).

Although there often is a greater chance of observing short-distance dispersal than dispersal over longer distances (Stenseth and Lidicker 1992), few marked Allegheny woodrats were recovered through trapping rock sites at the periphery of our study area. This suggests that disappearances of individuals are most likely due to mortality rather than dispersal (Jones 1987). If recapture data reflect survivorship, then it can be inferred that most woodrats in both juvenile and adult age classes perish within a given year.

In summary, available evidence suggests that in Allegheny woodrats, long-distance dispersal is uncommon and risky. Rather than dispersing as juveniles, individuals that do engage in natal dispersal wait until their first winter or the following spring. Males are more likely to travel long distances than females, but either sex or any age seems capable of the kind of dispersal that would allow recolonization of vacant habitat.

Survival

Maximum reported age for an Allegheny woodrat is 48 months in captivity (Poole 1940) and 58 months in the wild in West Virginia (Mengak et al. 2002). In our studies of a wild population in north-central West Virginia (Myers 1997, Manjerovic 2004, J. Zuck and P. Wood unpublished data), two adult females were recaptured 47 months after their initial capture as adults; thus, these individuals were approximately five years of age at their final capture. The maximum age at recapture for an adult male was 38 months after its initial capture as an adult making this individual approximately four years of age. The interval between initial and final capture was 2 days–38 months for 180 adult males ($\overline{X} = 8.5$ months) and 2 days–48 months for 196 adult females ($\overline{X} = 10.2$ months). When recaptures within a year are excluded, mean recapture interval was 16 months for males and 15 months for females. This suggests an average lifespan for Allegheny woodrats as slightly greater than 24 months in the wild for both sexes for individuals that survive to adulthood.

Survival estimates are difficult to quantify for Allegheny woodrats. Recapture rates are the most widely available estimate for annual survival but are heavily influenced by trapping effort and spatial orientation of rock outcrops. When rock sites occur in a long, contiguous pattern such as along the rim of a gorge, woodrat movement is generally linear, following the available rock habitat. If portions of some rock sites are inaccessible to trapping, marked woodrats foraging in those areas during live-trapping are unlikely to be captured. Rock sites that occur not only on the rim of a gorge but also at numerous interior locations create a multiple island effect where woodrats could move back and forth between rock sites. Simultaneous trapping of adjacent rock sites would enhance recapture potential. While it is possible that certain individuals may avoid traps after initial capture, many woodrats are recaptured the next night, often in the same trap. Low recapture rates could result from low, sporadic or unequal trapping effort, trap avoidance, dispersal or movement of individuals, or mortality. Thus, recapture rates would represent minimum survival rates and are most likely to be fairly accurate for adult females due to strong philopatry to specific rock sites and small home range sizes.

Over-winter recapture rates reported for juveniles in different parts of the range are somewhat variable. During 1995–2004 in north-central West Virginia, over-winter recapture rate was 25.6% (23 of 90 individuals) for juvenile females and 17.7% (14 of 79) for juvenile males (Myers 1997, Manjerovic 2004, J. Zuck and P. Wood unpublished data). In Virginia, the over-winter recapture rate for juveniles was 19%, with a lower rate for males (10%) than for females (29%) (Mengak 2002). It is not certain whether low recapture rates are due to dispersal or mortality. Juvenile movements of up to 3.3 km indicate that they are capable of dispersing a considerable distance, but few documented long-distance movements for juveniles suggest that mortality rates are high. For radiocollared juveniles, 4 of 8 females and 4 of 9 males were known or presumed dead within 12 weeks of being radiocollared (J. Zuck and P. Wood unpublished data). No marked individuals were recovered during extensive trapping of rock sites at the periphery of our north-central West Virginia study area further suggesting that most disappearances can be attributed to mortality rather than dispersal. At two sites in Pennsylvania, estimated juvenile survival was 61% for females and 25% for males, although these estimates may be high, particularly for females (J. Wright unpublished data). Generally higher survival and recapture rates for juvenile females combined with males being the more likely dispersers, suggests high juvenile mortality during dispersal.

Over-winter recapture rates were somewhat higher for adult Allegheny woodrats than juveniles in our north-central West Virginia study area (Myers 1997, Manjerovic 2004, J. Zuck and P. Wood unpublished data). During 1995–2004, 21.9% of 155 adult males were recaptured while 32.5% of 160 adult females were recaptured. Dispersal likely accounts for a low portion of disappearances of adult females because no long-distance movements have been documented for adult females in our study. For males, their lower overall recapture rate and their higher recapture rate at other outcrops suggests that males are more actively dispersing and moving through the population. Hassinger et al. (1996) had an overall 16% recapture rate for Allegheny woodrats in Pennsylvania. Other woodrat species also tend to have low recapture rates: *N. floridana* (4.3%, Fitch and Rainey 1956; 5%, Goertz 1970) and adult *N. macrotis* (38% recaptures of females and 27% of males, Matocq 2004).

Generally, low recapture rates for woodrat populations suggest high turnover rates. Between 1999 and 2000, five of 24 radiocollared Allegheny woodrats were killed by predators and six others disappeared with unknown fates (J. Zuck and P. Wood unpublished data). In Pennsylvania, estimated survival was 26.1% and 16.7% for one- and two-year old females, respectively, with somewhat higher survival rates of 38.1% and 25.0% for one- and two-year old males, respectively (J. Wright unpublished data). Whether survival data from these populations represent survival of normal or declining populations is not known. In New York and New Jersey, LoGiudice (2003) documented average survival after experimental release to 15.5 weeks at two sites highly contaminated by the raccoon roundworm (*Baylisascaris procyonis*); all individuals disappeared by 52 weeks after release. Approximately 15% of individuals survived to 52 weeks at two less contaminated sites with an average survival of 33.6 weeks post release.

Reproductive Parameters

The breeding season for Allegheny woodrats generally spans early spring until mid-fall. Reproduction has been reported in mid-March to mid-October in Pennsylvania (Poole 1940), early February–October in Virginia (Mengak 2002), and January–August in West Virginia (Myers 1997, J. Zuck and P. Wood unpublished data). The majority of reproduction occurs in early spring and late summer. Peak juvenile captures in Virginia (Mengak 2002) occurred in May–June (48%) with a smaller peak in September–October (26%). Based on weight at capture, most juveniles in north-central West Virginia appeared to be born in March and April (J. Zuck and P. Wood unpublished data). Myers (1997) estimated peak birth dates as early April. Although there seem to be reproductive peaks, estimated birth dates in January (J. Zuck and P. Wood unpublished data), juveniles observed in early February (Mengak 2002), and lactating females captured in October in Virginia (Mengak 2002) and West Virginia all suggest that reproduction may occur year round. Given field observations and reproductive patterns of other rodent species, woodrats most likely are reproductive opportunists that breed continuously when conditions allow. The high degree of variability in the length of the reproductive season is thought to be influenced by variability in mast crops, severity of winter (Whitaker and Hamilton 1998), and availability of security cover (Mengak 2002) that would lower the risk of predation and ameliorate some weather effects.

From studies of captive Allegheny woodrats, gestation is estimated to be 30–36 days, sexual maturity occurs at 3–4 months, and individuals may give birth at 10 months (Poole 1940). Poole (1940) suggested that they can raise two litters each year. The potential for multiple litters in the wild is supported by the bimodal peaks of juvenile captures in Virginia (Mengak 2002), captures of juveniles during each month in May–October in West Virginia, and the long length of the breeding season. Female *N. cinerea* in the wild produced up to two successful litters per year (Topping and Millar 1998).

Typical litter size from both field and captive studies is estimated to be 1–4 pups with 2–3 being most common. Although Bailey (1946; cited in Mengak 2002) reported litter size as 2–6 in Virginia, he presented no actual data. In Virginia, Mengak (2002) reported 2–3 pups in six litters ($\overline{X} = 2.3$). Myers (1997) reported one litter with two pups and two litters with four pups ($\overline{X} = 3.3$) in West Virginia. Poole (1940) reported 1–3 ($\overline{X} = 2.0$) pups in 10 captive litters. In captive individuals at the Baltimore Zoo, typical litter size was 2–3 with one litter of five pups (cited in Mengak 2002), while a laboratory born litter in Ohio had four pups (Kinsey 1977). Using genetic analyses to determine parentage, Manjerovic (2004) estimated that litter size for 15 wild females in West Virginia was 1–3 pups ($\overline{X} = 1.5$). Estimates of litter size for her study were based on captured juveniles, thus represent juveniles that survived beyond weaning age and are likely to be somewhat lower than the actual litter size born to females. Manjerovic (2004) also found half-sibling relationships among some juveniles suggesting possible multiple male paternity, which has been demonstrated in other rodents (e.g., Baker et al. 1999, Ohnishi et al. 2000).

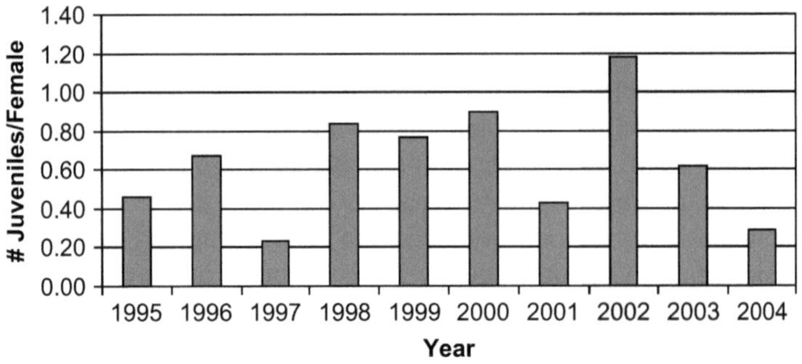

Fig. 3.2 Number of juvenile Allegheny woodrats (*Neotoma magister*) captured per adult female in north-central West Virginia during 1995–2004

Female Allegheny woodrats appear capable of producing offspring until at least five years of age based on mark-recapture data from north-central West Virginia (Myers 1997, Manjerovic 2004, J. Zuck and P. Wood unpublished data). At our study sites, two females, both approximately five years old, were lactating when captured. An additional four females that were at least four years of age all showed evidence of being reproductive. For adult females of all ages in our study, the proportion of female captures that showed evidence of reproduction varied across years and ranged from 62% to 100%. Topping and Millar (1998) determined that 75–89% of female *N. cinerea* successfully produced offspring in a given year, while Matocq (2004) found that 61–62% of female *N. macrotis* produced offspring.

Reproductive success is difficult to measure in a species such as the Allegheny woodrat that places its nests deep within rock outcrops. A surrogate measure of reproductive success is the ratio of juveniles captured per adult female (Lee and Tietje 2005). Capture ratios for *N. fuscipes* ranged from 0.3 to 1.0 juveniles per adult female in a declining population (Lee and Tietje 2005). In north-central West Virginia between 1995 and 2004, the ratio of juveniles to adult females captured varied among years from 0.24 to 1.18 juveniles per female ($\overline{X} = 0.68$; Fig. 3.2). This rate appears to be below replacement levels as this Allegheny woodrat population is showing declines (see below). At a site in Pennsylvania, 0.20–2.00 juveniles per female ($\overline{X} = 0.88$) were captured (J. Wright unpublished data). In southern West Virginia, at presumably poor quality sites on reclaimed surface mines, Chamblin, et al. (2004) captured juveniles at four of seven sites where females were captured; these sites had a mean capture ratio of 0.88 juveniles per female.

Population Trends and Factors Affecting Demography and Survival

Allegheny woodrat populations have been declining in recent decades over much of their range, although they were thought to be relatively stable in the core range in West Virginia. We conducted a mark-recapture study on a woodrat population in north-central West Virginia to gain a better understanding of the population's status (Myers 1997, Manjerovic 2004, J. Zuck and P. Wood unpublished data). During 1995–2003, we captured 439 individual woodrats (146 adult males, 170 adult females, 58 juvenile males, 65 juvenile females) over a total of 7644 trap nights at Cooper's Rock State Forest (CR). During 1999–2003 at Snake Hill Wildlife Management Area (SH) sites across the Cheat River from CR, we trapped for 1685.5 trap nights and captured 83 individual woodrats (33 adult males, 34 adult females, 9 juvenile males, 7 juvenile females). Significant ($p = 0.04$) declines occurred in adult female capture rates at CR but not in adult males and juveniles (Fig. 3.3;

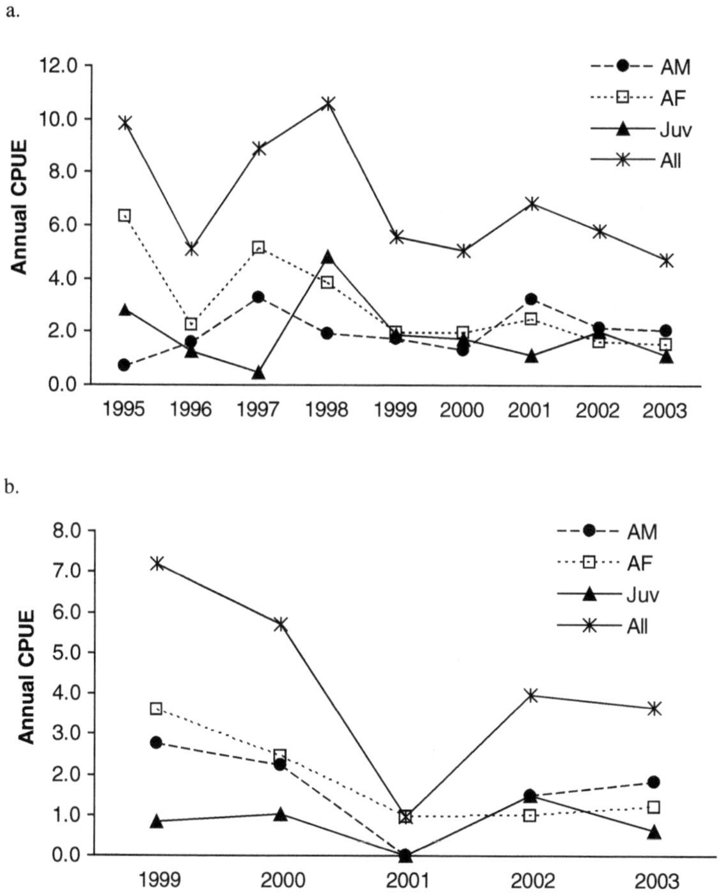

Fig. 3.3 Annual trends in captures adjusted per 100 trap nights (CPUE) for adult male (AM), adult female (AF), and juvenile (Juv) Allegheny woodrats (*Neotoma magister*) between 1999 and 2003 at (a) Coopers Rock State Forest (AM: $r = 0.35, p = 0.35$; AF: $r = -0.75, p = 0.02$; Juv: $r = -0.22, p = 0.56$; All: $r = -0.58, p = 0.10$) and (b) Snake Hill Wildlife Management Area (AM: $r = -0.53, p = 0.36$; AF: $r = -0.87, p = 0.06$; Juv: $r = -0.17, p = 0.79$; All: $r = -0.70, p = 0.19$) in West Virginia. Trends examined with Pearson product–moment correlation. Adapted from Manjerovic (2004)

Manjerovic 2004). Similarly, declines in female captures at SH approached significance ($p = 0.06$).

Rodent fecundity is often directly related to food availability; years of abundant mast production extend the breeding season and increase populations in subsequent years for some species (Smyth 1966, Gashwiler 1979, Wolff 1996). Because mast is a significant component of woodrat diets (Castleberry et al. 2002a), the reduction of mast crops has been suggested as a potential threat to the Allegheny woodrat (Balcom and Yahner 1996). The complete extirpation of northern populations also suggests that severe winters may compound the intensity of threats (Beans 1992). The raccoon roundworm may be a significant factor in declines of some northern populations (LoGiudice 2006, Chapter 2). In West Virginia, however, Owen et al. (2004) found no evidence of *B. procyonis*, so we were able to examine population impacts related only to mast availability and weather at our West Virginia study sites (Manjerovic 2004).

Regional mast production data were available from yearly surveys conducted by the West Virginia Division of Natural Resources. Because abundance of hard mast from preceding years is significantly related to densities of rodents in subsequent years (Wolff 1996, McShea 2000), we compared hard mast production from the previous year to annual woodrat capture rates. Soft mast is most abundant during spring and summer (Castleberry et al. 2002a), so we compared these values with captures from the current year. We found no significant correlations between yearly capture rates of adult male, adult female, or juvenile woodrats and indices of hard and soft mast abundance ($p = 0.17$–0.91; Manjerovic 2004). Other woodrat species have demonstrated opportunistic foraging behavior in which diet selection shifts in response to food availability and habitat change (McMurry et al. 1993). It is possible that the generalist diet and food caching behaviors of the Allegheny woodrat help in enhancing survival during winter months in years of lower mast production (Castleberry et al. 2002a). However, an increase in foraging distances following a poor mast year (Castleberry et al. 2001a) could make individuals more vulnerable to predation and lower the probability of survival.

We related annual woodrat capture rates to annual weather data from the Coopers Rock Weather Station. We calculated total snowfall and rainfall and average temperature (mean, maximum, and minimum) for each month, then averaged these for each season (Manjerovic 2004). Adult female capture rates decreased with an increase in summer rainfall and adult male capture rates were lower with low summer temperatures (Fig. 3.4). Snowfall in spring and lower winter temperatures were related to lower juvenile captures (Fig. 3.5) suggesting that juveniles born in late winter or early spring may experience lower survival in years with lower temperatures or with spring snowfall. Alternatively, the impacts on juveniles may be a result of winter weather affecting adult reproduction with females producing fewer offspring after a severe winter.

In *N. floridana*, security cover was the most important limiting factor to survival (McMurry et al. 1993). For Allegheny woodrats, outcrops probably provide better protection than stick nests but individuals must range farther for foraging which exposes them to greater risk of predation. Allegheny woodrats generally have larger

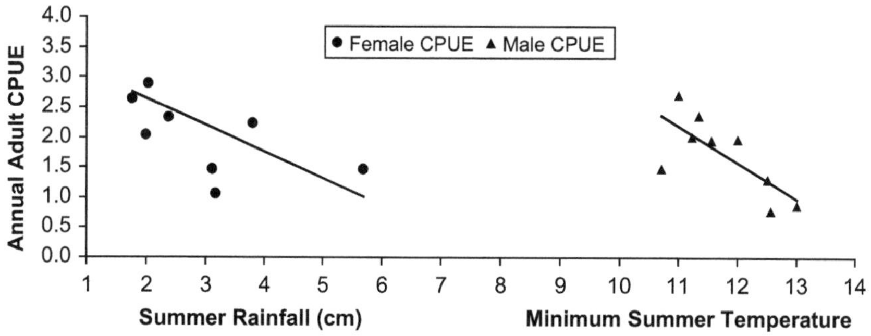

Fig. 3.4 Annual captures of adult female Allegheny woodrats (*Neotoma magister*) relative to summer rainfall (cm; $r = -0.75$, $p = 0.02$) and annual captures of adult males relative to minimum summer temperature (°C; $r = -0.80$, $p = 0.01$) in north-central West Virginia. Captures are adjusted per 100 trap nights (CPUE). Adapted from Manjerovic (2004)

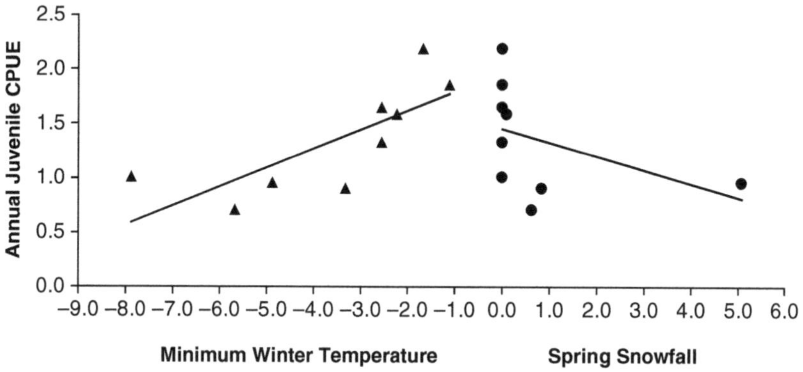

Fig. 3.5 Annual captures of juvenile Allegheny woodrats (*Neotoma magister*) in north-central West Virginia relative to minimum winter temperature (°C; $r = -0.85$, $p = 0.004$) and average spring snowfall (cm; $r = -0.67$, $p = 0.05$). Captures are adjusted per 100 trap nights (CPUE). Values plotted are untransformed data. Adapted from Manjerovic (2004)

home ranges than most woodrats (Castleberry et al. 2001a) except for *N. cinerea* which also use rock outcrops (Topping and Millar 1996).

Mengak (2002) suggested that low reproductive rates and specialized habitat requirements may make Allegheny woodrats vulnerable to population declines resulting from human activities, such as changes to forest composition (Balcom and Yahner 1996). Clearcutting and two-age regeneration harvesting (removing all trees except 6–10 residual trees/ha) in central West Virginia appear to have minimal direct impact if the forest overstory is retained immediately surrounding outcrops and intact forest is maintained on one adjacent side (Castleberry et al. 2001a). Limited timber harvesting near outcrops can benefit woodrats by increasing production of herbaceous vegetation and soft mast (Castleberry et al. 2001b) which are important foods during spring and summer (Castleberry et al. 2002a).

Limited dispersal based on field observations and few effective migrants per generation based on genetic data suggest that this species has low potential for recolonization of sites where subpopulations have become extinct. This low recolonization potential in turn makes this species highly sensitive to population and habitat perturbations. LoGiudice (2006) suggested that multiple stressors, all attributable to human activities, have contributed to collapse of Allegheny woodrat populations in northern portions of the range.

Conservation and Management of Declining Species

As a general approach for declining species, it is important to examine whether a causal factor in the decline is decreased reproduction, increased mortality, or a combination of these factors: in short, to identify extinction risk. It is also important to examine independent, possibly causative, variables (e.g., food resources, weather, disturbance factors, parasitism, etc.) and to relate these variables to population and demographic parameters. If these data cannot be collected during the course of a study, coordinating with agencies that have these data is imperative. An Allegheny woodrat population in north-central West Virginia with a mean ratio of 0.68 juveniles per adult female appears to be declining, a result that suggests this ratio is below replacement levels for the population. Reasons for the decline are not definitive, but winter weather is correlated with juvenile capture rates.

Declining species often exhibit a metapopulation structure due to loss and fragmentation of habitat. Maintaining linkages among subpopulations is critical for maintaining gene flow and recolonization through dispersal. It is important to understand the mechanics of natural dispersal (who disperses, when, who succeeds) to effectively manage a metapopulation. Barriers to dispersal must be identified and mitigated as they can destabilize metapopulations and limit recolonization of habitat patches.

For Allegheny woodrats, rivers are not a barrier to movements based on genetic data (Castleberry et al. 2002b, Manjerovic 2004) but highways, urbanization, and agricultural lands appear to be barriers (LoGiudice 2006). The habitat matrix surrounding Allegheny woodrat subpopulations can influence dispersal movements and hence gene flow. Some forms of anthropogenic disturbance in the surrounding matrix may act as barriers, reducing successful dispersal and isolating subpopulations. Species with a high degree of habitat specificity, such as the Allegheny woodrat, are especially vulnerable to anthropogenic disturbance (Rabinowitz et al. 1986). Although habitat disturbances associated with forest management could affect movements of Allegheny woodrats, Castleberry et al. (2001b) detected woodrats moving through clearcuts suggesting that low levels of forest management in a surrounding landscape do not limit movements. Castleberry (2000) concluded that "from a metapopulation perspective, it is unlikely that forest management has a significant effect on dispersal among rock outcrops and subsequently on gene flow." There is some evidence, however, that rock outcrops isolated by human disturbance may have a reduced likelihood of recolonization. Castleberry et al.

(2001b) captured no woodrats at outcrops where the overstory trees had recently been removed, although old latrines and middens were observed, indicating that woodrats were present before harvest. Anthropogenic disturbance in the form of outdoor recreational activities, such as hiking and climbing, appear to have variable effects on woodrat populations (Wood 2001) and likely depend on duration, intensity, and scale of the activity. Woodrats disappeared entirely from a site with high intensity rock climbing activity (M. Varner pers. comm.).

Management for maintaining genetic diversity of habitat specialists requires knowledge of their geographic distribution, their genetic variation, and the geographic scale at which populations function independently (Pope et al. 1996, Perault et al. 1997). On the basis of genetic data, Castleberry et al. (2002b) suggested that the functional unit for Allegheny woodrats may be inhabitants of individual rock outcrops. Manjerovic (2004) found that genetic and spatial distances were positively correlated for individual outcrops even though all outcrops were not significantly differentiated. The lowest effective migration rate was between subpopulations separated by the greatest distance (Castleberry et al. 2002b, Manjerovic 2004). Thus, geographic distance can be a meaningful barrier to dispersal, and management actions need to conserve populations that are close enough together to permit sufficient genetic exchange.

Occasional dispersal is a critical feature of Allegheny woodrat metapopulation persistence, but we still know very little about the conditions that induce an individual Allegheny woodrat to disperse to a new habitat patch. Male Allegheny woodrats are more likely to travel long distances than females, but all sex and age classes are capable of long-distance movements that would allow recolonization of vacant habitat. Still, natal and breeding dispersal appear to be uncommon with high mortality of dispersers. Rather than dispersing as juveniles, individuals that do engage in natal dispersal wait until their first winter or the following spring. Thus, natal dispersal in this species does not appear to conform to conventional wisdom that dispersal occurs in juveniles, which cautions us to avoid making assumptions based on norms for other species.

Acknowledgments Funding and other support to complete our studies were provided by the West Virginia Division of Natural Resources Wildlife Diversity Program, the Division of Forestry and Natural Resources at West Virginia University, Clarion University, and the U.S. Geological Survey's Cooperative Research Units Program and West Virginia Cooperative Fish and Wildlife Research Unit. A number of graduate and undergraduate students at West Virginia University contributed to data collection, in particular, Ryan Myers, Jon Zuck, Mary Beth Manjerovic, and Matthew Schnupp. Helpful comments on earlier drafts were provided by Joe Duchamp, Mary Beth Manjerovic, John D. Peles, Janet Wright, Jon Zuck, and an anonymous reviewer.

References

Bailey, J. W. 1946. The Mammals of Virginia. Williams Printing Company, Richmond, Virginia.
Baker, R. J., Makova, K. D., and Chesser, R. K. 1999. Microsatellites indicate a high frequency of multiple paternity in *Apodemus* (Rodentia). Molecular Ecology 8:107–111.

Balcom, B. J., and Yahner, R. H. 1996. Microhabitat and landscape characteristics associated with the threatened Allegheny woodrat. Conservation Biology 10:515–525.

Beans, B. E. 1992. Without a trace: the puzzling demise of the Allegheny woodrat. Audubon 92:32–34.

Boonstra, R., Krebs, C. J., Gaines, M. S., Johnson, M. L., and Craine, I. T. 1987. Natal philopatry and breeding systems in voles (*Microtus* spp.). Journal of Animal Ecology 56:655–673.

Castleberry, S. B. 2000. Conservation and management of the Allegheny woodrat in the central Appalachians. Ph.D. Dissertation, West Virginia University, Morgantown, West Virginia.

Castleberry, S. B., Ford, W. M., Wood, P. B., Castleberry, N. L., and Mengak, M. T. 2001a. Movements of Allegheny woodrats in relation to timber harvesting. Journal of Wildlife Management 65:148–156.

Castleberry, S.B., Wood, P. B., Ford, W. M., Castleberry, N. L., and Mengak, M. T. 2001b. Summer microhabitat selection by foraging Allegheny woodrats in a managed forest. American Midland Naturalist 147:93–101.

Castleberry, N. L., Castleberry, S. B., Ford, W. M., Wood, P. B., and Mengak, M. T. 2002a. Allegheny woodrat (*Neotoma magister*) food habits in the central Appalachians. American Midland Naturalist 147:80–92.

Castleberry, S. B., King, T. L., Wood, P. B., and Ford, W. M. 2002b. Microsatellite DNA analysis of population structure in Allegheny Woodrats (*Neotoma magister*). Journal of Mammalogy 83:1058–1070.

Chamblin, H. D., Wood, P. B., and Edwards, J. W. 2004. Allegheny woodrat (*Neotoma magister*) use of rock drainage channels on reclaimed mines in southern West Virginia. American Midland Naturalist 151:346–354.

Conditt, S. A., and Ribble, D. O. 1997. Social organization of *Neotoma micropus*, the southern plains woodrat. American Midland Naturalist 137:290–297.

Cranford, J. A. 1977. Home range and habitat utilization by *Neotoma fuscipes* as determined by radiotelemetry. Journal of Mammalogy 58:165–172.

Diffendorfer, J. E., Gaines, M. S., Slade, N. A., and Holt, R. D. 1999. Patterns and impacts of movements at different scales on small mammals. *In* Landscape Ecology of Small Mammals, eds. G. W. Barrett and J. D. Peles, pp. 63–88. Springer-Verlag, New York.

Fitch, H. S., and Rainey, D. G. 1956. Ecological observations of the woodrat, *Neotoma floridana*. University of Kansas Publications, Museum of Natural History 8: 499–533.

Gashwiler, J. S. 1979. Deer mouse reproduction and its relation to the tree seed crop. American Midland Naturalist 102:95–104.

Goertz, J. W. 1970. An ecological study of *Neotoma floridana* in Oklahoma. Journal of Mammalogy 51:94–104.

Greenwood, P. J. 1980. Mating systems, philopatry and dispersal in birds and mammals. Animal Behavior 28:1140–1162.

Hassinger, J. D., Butchkoski, C. M., and Diefenbach, D. R. 1996. Fragmentation effects on the occupancy of forested Allegheny woodrat (*Neotoma magister*) colony area. Research report to the Pennsylvania Game Commission, Harrisburg, Pennsylvania.

Howard, W. E. 1960. Innate and environmental dispersal of individual vertebrates. American Midland Naturalist 63:152–161.

Jones, W. T. 1987. Dispersal patterns in kangaroo rats (*Dipodomys spectabilis*). *In* Mammalian Dispersal Patterns: The Effects of Social Structure on Population Genetics, eds. B. D. Chepko-Sade and Z. T. Halpin, pp. 119–127. University of Chicago Press, Chicago, Illinois.

Kinsey, K. P. 1977. Agonistic behavior and social organization in a reproductive population of Allegheny woodrats, *Neotoma floridana magister.* Journal of Mammalogy 58:417–419.

Lee, D. E., and Tietje, W. D. 2005. Dusky-footed woodrat demography and prescribed fire in a California oak woodland. Journal of Wildlife Management. 69:1211–1220.

Lidicker, W. Z., Jr. 1975. The role of dispersal in the demography of small mammals. *In* Small Mammals: Their Productivity and Population Dynamics, eds. F. B. Golley, K. Petrusewicz and L. Ryskowski, pp. 103–128. Cambridge University Press, Cambridge, Massachusetts.

Linsdale, J. M., and Tevis, L P., Jr. 1951. The Dusky-Footed Woodrat. University of California Press, Berkeley, California.

LoGiudice, K. 2003. Trophically transmitted parasites and the conservation of small populations: raccoon roundworm and the imperiled Allegheny woodrat. Conservation Biology 17:258–266.

LoGiudice, K. 2006. Toward a synthetic view of extinction: a history lesson from a North American rodent. BioScience 56:687–693.

Lynch, M. F., Fresnock, A. L., and Van Vuren, D. 1994. Home range and social structure of the dusky-footed woodrat (*Neotoma fuscipes*). Northwestern Naturalist 75:73–75.

Manjerovic, M. E. 2004. Demography and genetic structure of an Allegheny woodrat population in northcentral West Virginia. M.S. Thesis. West Virginia University, Morgantown, West Virginia.

Matocq, M. D. 2004. Reproductive success and effective population size in woodrats (*Neotoma macrotis*). Molecular Ecology13: 1635–1642.

McCullough, D. R. 1996. Metapopulations and Wildlife Conservation. Island Press, Washington, D.C.

McMurry, S. T., Lochmiller, R. L., Boggs, J. F., Leslie, D. M., Jr., and Engle, D. M. 1993. Opportunistic foraging of eastern woodrats (*Neotoma floridana*) in manipulated habitats. American Midland Naturalist 130:323–337.

McShea, W. J. 2000. The influence of acorn crops on annual variation in rodent and bird populations. Ecology 81:228–238.

McShea, W. J., Madison, D. M. 1992. Appendix 2. Alternative approaches to the study of small mammal dispersal: insights from radio telemetry. *In* Animal Dispersal: Small Mammals as a Model, eds. N. C. Stenseth and W. C. Lidicker, Jr., pp.319–332. Chapman and Hall, New York.

Mengak, M. T. 2002. Reproduction, juvenile growth and recapture rates of Allegheny woodrats (*Neotoma magister*) in Virginia. American Midland Naturalist 148:155–162.

Mengak, M. T., Castleberry, S. B., Ford, W. M., Rodrigue, J., and Castleberry, N. L. 2002. Longevity record for a wild Allegheny woodrat (*Neotoma magister*) in West Virginia. Virginia Journal of Science 53:167–170.

Moses, R. A., and Millar, J. S. 1994. Philopatry and mother-daughter associations in bushy-tailed woodrats: space use and reproductive success. Behavioral Ecology and Sociobiology 35: 131–140.

Myers, R. T. 1997. Microhabitat and ecology of the Allegheny woodrat in northcentral West Virginia. M.S. Thesis. West Virginia University, Morgantown, West Virginia.

Newcombe, C. L. 1930. An ecological study of the Allegheny cliff rat (*Neotoma pennsylvanica* Stone). Journal of Mammalogy 11:204–211.

Ohnishi, N., Saitoh, T., and Ishibashi, Y. 2000. Spatial genetic relationships in a population of the Japanese wood mouse *Apodemus argenteus*. Ecological Research 15:285–292.

Owen, S. F., Edwards, J. W., Ford, W. M. Crum, J. M., and Wood, P. B. 2004. Raccoon roundworm in raccoons in central West Virginia. Northeastern Naturalist 11:137–142.

Pearson, P. G. 1952. Observations concerning the life history and ecology of the woodrat, *Neotoma floridana floridana* (Ord). Journal of Mammalogy 33:459–463.

Perault, D. R., Wolf, P. G., and Edwards, T. C., Jr. 1997. Hierarchical analysis of genetic partitioning by *Tamias minimus* and *T. umbrinus*. Journal of Mammalogy 78:134–145.

Poole, E. L. 1940. A life history sketch of the Allegheny woodrat. Journal of Mammalogy 21: 249–270.

Pope, L. C., Sharp, A., and Moritz, C. 1996. Population structure of the yellow-footed rock-wallaby *Petrogale xanthopus* (Gray, 1854) inferred from mtDNA sequences and microsatellite loci. Molecular Ecology 5:629–640.

Rabinowitz, D., Cairns, S., and Dillon, T. 1986. Seven forms of rarity and their frequency in the flora of the British Isles. *In* Conservation Biology: The Science of Scarcity and Diversity, ed. M. E. Soulé, pp.182–204. Sinauer and Associates, Sunderland, Massachusetts.

Smyth, M. 1966. Winter breeding in woodland mice, *Apodemus sylvaticus*, and voles, *Clethrionomys glareolus*, and *Microtus agrestis*, near Oxford. Journal of Animal Ecology 35:471–485.

Stenseth, N. C., and Lidicker, W. Z., Jr. (eds.). 1992. Animal Dispersal: Small Mammals as a Model. Chapman and Hall, New York.

Topping, M. G., and Millar, J. S. 1996. Foraging movements of female bushy-tailed woodrats (*Neotoma cinerea*). Canadian Journal of Zoology 74:798–801.

Topping, M. G., and Millar, J. S. 1998. Mating patterns and reproductive success in the bushy-tailed woodrat (*Neotoma cinerea*), as revealed by DNA fingerprinting. Behavioral Ecology and Sociobiology 43:115–124.

Waser, P. M., and Jones, W. T. 1983. Natal philopatry among solitary mammals. Quarterly Review of Biology 58:355–390.

Whitaker, J. O., Jr., and Hamilton, W. J., Jr. 1998. Mammals of the Eastern United States. Cornell University Press, Ithaca, New York.

Wiens, J. A. 1996. Wildlife in patchy environments: populations, mosaics, and management. *In* Metapopulations and Wildlife Conservation, ed. D. R. McCullough, pp.53–84. Island Press, Washington, D.C.

Wolff, J. O. 1996. Population fluctuations of mast-eating rodents are correlated with production of acorns. Journal of Mammalogy 77:850–856.

Wood, P. B. 2001. Characteristics of Allegheny woodrat (*Neotoma magister*) habitat in the New River Gorge National River, West Virginia. Final Project report to the National Park Service, New River Gorge National River, Glen Jean, West Virginia.

Wright, J. 1996. Space and resource use by Allegheny woodrats: foraging movements and scarce resources. Abstracts of the Allegheny woodrat Recovery Group Meeting, Ferrum College, Ferrum, Virginia.

Chapter 4
Home Range, Movements, and Habitat Selection

Steven B. Castleberry

Introduction

Home range, movements, and habitat selection are commonly used metrics to examine the response of vertebrates to habitat change. Localized habitat disturbance from land management activities, such as forest management, have been proposed as a threat to Allegheny woodrat (*Neotoma magister*) populations (Castleberry et al. 2001). Dependence upon rock outcrops makes Allegheny woodrats especially vulnerable to disturbance in habitat surrounding the outcrop. Local disturbances can change vegetative structure and resource distribution, and these have been shown to influence demographics, home range size, and microhabitat use of rodents (Yahner 1986, Bowers et al. 1990, Jones 1990, Lacher and Mares 1996, Taulman et al. 1998).

Few data are available on home range, movements, and habitat selection by the Allegheny woodrat. Furthermore, few studies have examined the effects of local disturbances on these aspects of woodrat ecology. In this chapter, I summarize the existing knowledge on home range, movements, and habitat use in general and how these variables are influenced by local disturbance, such as timber harvesting.

Home Range Size and Foraging Movements

Three published studies have examined home range size and movements of the Allegheny woodrat (Castleberry et al. 2001, Mengak 2002, Hornsby et al. 2005). When considered together, these three studies provide estimates of home range size and movements throughout the year from the east-central portion of the species' range (Table 4.1). Published home range estimates are not available from other areas within the distribution.

Castleberry et al. (2001) used radiotelemetry to examine home range size and foraging movements in the Allegheny Mountains and Plateau region of east-central

S.B. Castleberry
Warnell School of Forestry and Natural Resources, University of Georgia, Athens, GA 30602

J.D. Peles, J. Wright (eds.), *The Allegheny Woodrat*, DOI: 10.1007/978-0-387-36051-5_4,

Table 4.1 Comparison of mean (±SE) home range and foraging movements of male (M) and female (F) Allegheny woodrats (*Neotoma magister*) among seasons reported in three published studies. Methods for each study are described in the text

	Season								
	Summer[1]			Fall[2]			Winter[3]		
	M	F	M/F	M	F	M/F	M	F	M/F
Home range (ha)	6.5	2.2	–	0.49	0.78	0.65	0.11	0.18	–
	(1.8)	(1.3)		(0.13)	(0.35)	(0.20)		(0.003)	
Foraging rate (m/hr)	–	–	202.6	94.1	78.3	86.2	–	–	–
			(9.2)	(23.1)	(22.6)	(15.7)			
Maximum distance (m)	–	–	151.0	45.1	63.8	54.5	–	–	–
			(7.3)	(8.7)	(25.0)	(13.0)			
Linear distance (m) (Site 1)	–	–	–	–	–	–	169.7	190.6	–
							(108.3)	(92.3)	
Linear distance (m) (Site 2)	–	–	–	–	–	–	102.1	104.0	–
							(68.8)	(68.1)	

[1] Summarized from Castleberry et al. (2001); home range data pooled across sexes, years, and timber harvest methods; movement data pooled across years and timber harvest methods.
[2] Summarized from Hornsby et al. (2005); data pooled across timber harvest methods.
[3] Summarized from Mengak (2002).

West Virginia. They reported mean (±SE) home range estimates of Allegheny woodrats during summer (May–August) in east-central West Virginia as 6.5 (±1.8) ha and 2.2 (±0.3) ha for males (n = 19) and females (n = 18), respectively, and 4.4 (±1.0) ha for males and females combined. These home range estimates during summer were considerably larger than reported estimates from fall (Hornsby et al. 2005) and winter (Mengak 2002). On the same West Virginia study area using similar methods as Castleberry et al. (2001), Hornsby et al. (2005) reported mean home range size in fall (October–December) as 0.65 (±0.20) ha for males and females combined, and 0.49 (±0.13) ha and 0.78 (±0.35) ha for males (n = 3) and females (n = 4), respectively. Mengak (2002) used radiotelemetry and live-trapping to estimate mean home range size during winter in Virginia. Based on radiotelemetry, he reported mean home range size of four females at 0.18 (±0.003) ha and the home range of one male at 0.11 ha. Mean home ranges derived from trapping at two Virginia study sites were estimated at 0.19 (±0.14) and 0.23 (±0.20) for females (n = 14) and males (n = 15), respectively, and 0.07 (±0.08) and 0.06 (±0.05) ha for females (n = 10) and males (n = 9), respectively.

The larger home ranges in summer as compared to fall and winter are most likely explained by variation in movements of woodrats in response to seasonal variation in food availability. In spring and summer, woodrats forage primarily on fungi, green vegetation, and various fruits (Castleberry et al. 2002a, Castleberry and Castleberry Chapter 6) that may be located away from the rock outcrops. In late fall and winter, woodrats are relying primarily on cached foods and seldom move >50 m from the den (Hornsby et al. 2005). Although annual home range estimates are not available, they would likely be intermediate as the larger summer home range sizes are averaged with the smaller fall/winter home ranges.

Estimates of Allegheny woodrat home range, especially in summer, generally are larger than estimates reported for most other *Neotoma* species (Goertz 1970, Cranford 1977, Vaughan and Schwartz 1980, Lynch et al. 1994, Conditt and Ribble 1997). This disparity may be explained by *N. magister's* dependence on rocky habitats. Topping and Millar (1996a) reported home range size estimates for the bushy-tailed woodrat (*N. cinerea*), a species also associated with rock outcrops, obtained in spring and summer in Canadian Rockies. Their home range sizes were similar to those reported by Castleberry et al. (2001). Home ranges of bushy-tailed woodrats over a three-year period ranged from 5.10 (±0.84) ha to 7.27 (±0.70) for males and 3.13 (±0.65) to 4.44 (±0.33) for females. Topping and Millar (1996a) speculated that the relatively large home ranges they observed in the bushy-tailed woodrat were a consequence of dependence on rock outcrops for den sites. Outcrops may not be close to necessary food resources, so that individuals may need to forage at farther distances. Outcrops probably provide better protection from weather and predators and a more stable microclimate than exposed stick nests. Thus, the benefits of occupying a rock outcrop probably outweigh the energetic costs and possible predation risk associated with a large home range (Topping and Millar 1996a).

Movement patterns with regard to season appear to follow trends similar to home range size. Castleberry et al. (2001) and Hornsby et al. (2005) collected foraging movement data on the same study area using similar methods, with the primary difference between the studies being the season of data collection. In both studies, two indices of foraging movements were calculated from telemetry locations. Castleberry et al. (2001) reported mean foraging rate (defined as the sum of the linear distances between successive telemetry locations for each night, divided by the time elapsed between the first and last locations) of Allegheny woodrats in summer as 202.6 (±9.2) m/hr over all individuals combined. Maximum nightly distance traveled from the den (defined as the linear distance from the den to the farthest telemetry location for each night) averaged 151.0 (±7.3) m. In fall, Hornsby et al. (2005) reported estimates of mean foraging rate and maximum nightly distance traveled as 86.2 (±15.5) m/hr and 54.2 (±13.0) m, respectively, for all individuals. Mengak (2002) used trapping data to examine winter movements on two study sites in Virginia. He reported mean linear movement distances (calculated as the distance between the farthest two trap stations at which an animal was captured) as 169.7 m (±108.3) and 190.6 m (±92.3), and 102.1 (±68.8) and 104.0 (±68.1) for males and females, respectively.

The greater foraging movement distances in summer likely reflect the lack of food availability and reliance on cached food in fall and winter. Restricted movement away from the den due to the reliance on cached foods has similar effect on foraging movements as on home range. However, breeding in Allegheny woodrats begins in winter (Castleberry et al. 2006) and may influence movement patterns and rates. Mengak (2002) observed that movements determined through radiotelemetry appeared to increase in February with the onset of the breeding season.

There are limitations in using data from multiple studies for comparing seasonal home range size and movement distances among seasons. Differences in the data collection and analysis methods used in the three investigations may affect seasonal comparisons. Furthermore, to date, only one study from each season has been published. Nonetheless, results seem consistent with expectations based on woodrat life history. For example, home range size and movement distances obtained through trapping by Mengak (2002) in winter generally were consistent with those obtained by Hornsby et al. (2005) in fall using radiotelemetry. Similar home range and movements in fall and winter would be expected as woodrats rely on cached food throughout both the seasons. Castleberry et al. (2001) and Hornsby et al. (2005) used similar methods on the same study area to obtain home range sizes and movement distances in summer and fall, respectively. Thus, the differences observed likely are representative of actual seasonal differences rather than artifacts of methodological differences. However, additional studies of seasonal home range and movements are needed to validate previous results. Additional studies also are needed to investigate differences between sex and age classes, particularly as related to energetic requirements of breeding adult females (Cranford 1977, Topping and Millar 1996b).

Habitat Selection

Several investigations have provided a quantitative examination of characteristics of rock outcrops and surrounding areas associated with woodrat presence (Balcom and Yahner 1996, Myers 1997, Bommarito 1999, Krupa et al. 2004). Proportional use of available habitats within the home range (Castleberry et al. 2001) and characteristics of habitat at known foraging locations (Castleberry et al. 2002b) also have been studied. Other studies have noted use of various habitats by the species (Ivanovich et al. 2001, Wood 2001, Chamblin et al. 2004). In the following discussion I provide an overall synthesis of Allegheny woodrat habitat selection.

Several generalities regarding the association of habitat characteristics with Allegheny woodrat presence at rock habitats may be drawn from previous studies. The physical characteristics and size of the rock habitat are critical for woodrat presence. Myers (1997) and Krupa et al. (2004) concurred that larger rock outcrops with a moderate to high number of crevices are considerably more likely to be occupied. In both studies, >84% of all occupied sites exhibited these two basic characteristics. Myers (1997) suggested that larger outcrops with numerous crevices provide increased rock surface largely devoid of leaf litter that provide foraging pathways on which woodrats are less likely to be detected by predators. Chamblin et al. (2004) also noted the importance of rock surface to woodrat presence, suggesting that the area of rock surface reflected the complexity of rock crevices and the ability of a site to provide adequate protection from predators. A complex network of crevices is important in providing denning areas that are inaccessible to many predators (Newcombe 1930, Poole 1940). Although most studies have examined emergent rock outcrops as woodrat habitat, Ivanovich et al. (2001) examined the potential importance of caves as Allegheny woodrat habitat in Kentucky and found that susceptibility of the underlying geology to karstification (erosion of limestone to form caves) was an important variable in predicting occurrence.

Results of previous studies strongly suggest that increased probability of outcrop occupancy by woodrats is associated with increasing slope (Balcom and Yahner 1996, Myers 1997, Bommarito 1999, Ivanovich et al. 2001). Myers (1997) suggested that slope and important physical characteristics of outcrops are related. Outcrops on steep slopes are exposed to greater rates of erosion resulting in larger portions of the outcrop being exposed, increasing the rate of weathering. Increased weathering causes the rocks to break, crack, and fall. Breaking and cracking, along with the resulting accumulated fallen rock, produces numerous caves and crevices at the base of the outcrop that are important for woodrat presence. Bommarito (1999) suggested that the importance of steep slopes reflected the presence of cliffline habitats which were the only occupied rock formations on her Kentucky study area.

Most habitat studies have concluded that human disturbance in the proximity (generally <1000 m) of occupied outcrops does not influence Allegheny woodrat occupancy. Balcom and Yahner (1996) and Myers (1997) found no evidence to suggest that disturbance in the proximity of rock outcrops affected woodrat presence. Common types of disturbance in these studies were major roads, railroad grades, and right-of-way clearings. Wood (2001) found that disturbance at the site was the single best predictor of woodrat abundance at New River Gorge National River in south-central West Virginia. However, counter to the expected outcome, woodrats were more abundant where disturbance levels from recreational rock climbing were greater. While it is likely that characteristics making rock habitats suitable for recreational rock climbing also make them suitable for woodrat presence, these results suggest that Allegheny woodrats are somewhat tolerant of human presence. Castleberry et al. (2001) found that clearcutting directly adjacent to outcrops had little effect on woodrat movements and that woodrats commonly foraged in one- to three-year-old clearcuts. In Kentucky, Ivanovich et al. (2001) found that occupied outcrops were farther from human disturbance, especially large (100-ha) disturbance patches, than random locations. However, their data suggest that woodrat presence is most likely influenced by human disturbance at the landscape scale (i.e., farther distances from the outcrop), as suggested by Balcom and Yahner (1996) and Ford et al. (2006), rather than local-scale disturbance.

Although Balcom and Yahner (1996) and Myers (1997) speculated that disturbance near colonies may increase predation from edge-related predators, such as great-horned owls (*Bubo virginianus*), neither study demonstrated any influence of local disturbance on woodrat presence. Mengak (2002) captured Allegheny woodrats in an old field as well as in the adjacent intact forest, concluding that there were no negative effects of the habitat disturbance on movement patterns. The types of disturbances most commonly found in the proximity of rock outcrops in previous studies, such as right-of-ways and clearcuts, may contain adequate vegetative cover to provide protection from avian predators. Other types of disturbance that result in greater reductions in vegetative cover (e.g., urban development or agricultural development) may have negative consequences.

Studies examining the influence of overstory stand condition have demonstrated variable results. The apparently lack of congruency in results may be in part due to lack of standardization in methods used to quantify overstory condition. Wood (2001) examined tree density in stem size classes and found evidence of a positive relationship between 3.0 and 8.0 cm DBH tree density and woodrat abundance. Using similar methods, Myers (1997) found a negative relationship between woodrat presence and density of trees in the 0.0–15.0 cm DBH size class, but larger diameter trees were present at occupied as well as unoccupied sites. Bommarito (1999) found a positive relationship between woodrat presence and tree density, but did not examine stem size classes separately. Balcom and Yahner (1996) found a positive relationship between overstory basal area and woodrat presence, but did not examine tree density. The lack of standardization in describing overstory condition among the studies makes drawing definite conclusions difficult. Nonetheless, all studies indicate that some level of overstory cover at the rock outcrop is necessary for woodrat presence.

Although overstory tree cover appears to be an important factor influencing presence, woodrats appear to tolerate a wide range of canopy cover. Chamblin et al. (2004) found that woodrats occupying drainage channels associated with reclaimed coal mines in southern West Virginia were associated with lower levels of overstory tree cover (~50%) than most other studies. Although Castleberry et al. (2001) found no effect of canopy removal adjacent to outcrops from clearcutting on woodrat foraging movements, they noted that woodrats were not found in rock outcrops where the canopy directly over the outcrop had been removed. The increase in solar radiation resulting from complete canopy removal may create unsuitable thermal conditions in the exposed outcrop.

Castleberry et al. (2001) examined proportional use of clearcut and intact forest at clearcut sites where woodrats had access to the harvested area on one side of the outcrop and intact forest on the other. Selection of the two habitats did not differ from random, indicating that woodrats used clearcuts and adjacent forested areas in proportion to their availability. The authors suggested that use of clearcut areas in summer is facilitated by dense growth of ground-level vegetation and an increase in food resources following canopy removal. Previous studies on other small mammal species have demonstrated increases in abundance following clearcutting, presumably in response to increases in ground cover and food availability (Kirkland 1977, Monthey and Soutiere 1985, Clough 1987). Succulent growth from hardwood stump sprouts in clearcuts represents a more abundant and palatable food for woodrats than vegetation under the forest canopy. Although Castleberry et al. (2001) recommend harvest planning that retains intact forest immediately covering rock outcrops on one adjacent side, they suggest that resource availability can be sufficiently maintained, or possibly improved, by timber harvests in the proximity of occupied outcrops.

In the only study to examine foraging habitat selection, Castleberry et al. (2002b) examined microhabitat characteristics of Allegheny woodrat foraging locations in relation to timber harvesting practices. Individuals at clearcut, diameter-limited, and intact forest stands were radio-collared and located during nocturnal foraging bouts. Physical and vegetative characteristics were measured at the foraging locations and were compared to random locations within each individual's foraging range. Interestingly, the type of harvest method did not influence selection of foraging sites by woodrats. However, when foraging and random locations were compared across harvesting methods, it became apparent that woodrats selected foraging sites with more diverse understory vegetation, regardless of overstory stand condition. The authors concluded that Allegheny woodrats are capable of tolerating a broad range of macrohabitats created through anthropogenic disturbance, as evidenced by presence at clearcut, diameter-limit, and intact forest stands, but habitat selection is largely determined by microhabitat factors. They further suggested that selection at the microhabitat scale is likely explained by the high mobility of the species and its generalist herbivore diet. Allegheny woodrats are capable of searching larger areas to locate suitable foraging microhabitats than other small mammals.

Local Habitat Disturbance and Conservation of the Species

In general, Allegheny woodrats appear tolerant of local (generally defined as <1000 m from a site) disturbance as long as overstory tree cover is retained at the rock outcrop on at least one adjacent side (Castleberry et al. 2001). Several studies have examined the influence of local disturbance on presence or abundance, and all concluded that disturbance had little or no influence (Balcom and Yahner 1996, Myers 1997, Ford et al. 2006). In fact, one study noted a positive relationship between woodrat abundance and anthropogenic disturbance (Wood 2001). The few studies that examined woodrat home range and movements have suggested little influence of local disturbance. However, there are several factors to be considered regarding the relationship between Allegheny woodrats and local disturbance.

The influence of local disturbance on home range and movements appears to be variable, depending on seasonal and annual differences in food resources. Castleberry et al. (2001) noted that patterns in home range size and foraging movements were larger and more variable in the first year of their study and largest in stands that had been harvested using a diameter-limit approach. They speculated that the patterns were related to differences in hard mast, primarily acorn, abundance. They only observed differences in home range and foraging movements in the year following a poor mast crop, and observed the greatest increase at the diameter-limit harvest sites where mast producing trees were scarce and widely scattered and where understory vegetation was limited. Low mast abundance and low understory diversity may have required woodrats to travel farther distances while foraging, subsequently increasing home range size. In fall, Hornsby et al. (2005) reported no differences in home range and movements between harvested and unharvested stands. The authors suggested that woodrats are affected less by timber harvesting in fall and winter when movements away from the outcrop are restricted.

Studies that have examined canopy cover in relation to the presence of woodrats have documented that a minimum level of canopy cover directly over the rock outcrop site is critical for woodrat presence. Most have found that occupied sites have canopy cover >71.5% (Myers 1997, Bommarito 1999). However, Chamblin et al. (2004) found that woodrats occupied rocky drainage channels with a mean canopy cover of 47.9%, ranging from 10% to 100%. Castleberry et al. (2001) noted that woodrats were not found in rock outcrops where the canopy directly over the outcrop had been removed. Interestingly, no studies that have quantitatively examined overstory condition relative to woodrat presence have found canopy cover to be an important variable in predicting presence, suggesting that above a minimum threshold level woodrats tolerate a broad range of canopy cover.

Although canopy cover at the rock outcrop is important for woodrat presence, canopy cover in foraging areas appears less important. Several studies noted disturbances such as roads, railroad and utility right-of-ways, and clearcuts adjacent to colony sites (Balcom and Yahner 1996, Myers 1997, Castleberry et al. 2001, 2002b). It is likely that these types of disturbances provide foraging opportunities as well as adequate cover for protection from predators (Castleberry et al. 2001). Similarly, Sakai and Noon (1997) concluded that dusky-footed woodrats (*N. fuscipes*) in northwestern California apparently were less susceptible to predation by northern spotted owls (*Strix occidentalis caurina*) when foraging in young, dense habitats than in more open, older-aged forests.

The influence of local habitat disturbance on gene flow among colonies, and subsequently on population structure, is unknown. Studies on movements and habitat use (Castleberry et al. 2001, 2002b) have shown that Allegheny woodrats commonly cross roads, right-of-ways, and harvested stands, suggesting that these types of disturbances do not severely impede dispersal among geographically proximate colonies. Larger scale disturbances and habitat modifications, such as intensive agriculture and residential development, appear to have a greater influence on population structure (Balcom and Yahner 1996, Ford et al. 2006). These disturbance types result in large-scale habitat fragmentation that may isolate rock outcrops and reduce recolonization probabilities (Ford et al. 2006, LoGiudice 2006). LoGiudice (2006) speculated that impeded dispersal as a result of habitat fragmentation was responsible for extirpation of the species in New York.

Although the body of literature suggests that local disturbance is not a major factor contributing to the decline of the species, most of the available data have been gathered from within the core of the distribution where populations appear relatively stable. The influence of local disturbance is not well studied in the northern periphery of the distribution where population declines are most pronounced. LoGiudice (2006) suggested that habitat disturbance has played a role, acting in concert with several other factors, in the observed population declines in those areas.

Conservation and Management of Declining Species

In this chapter, I have provided an overview of home range, movements, and habitat use by Allegheny woodrats. Additionally, I have summarized the available literature on the response of the species to local habitat disturbance. Using home range and movements as metrics for examining the response of a species to habitat change has general applicability to other declining species. However, some factors should be taken into consideration when relating inferences drawn from the Allegheny woodrat to other species.

The foraging strategy of the Allegheny woodrat may make inferences particularly pertinent to other central-place foraging species. The typical assumption in habitat use studies is that an individual's selection of a habitat is only influenced by the habitat type (Rosenberg and McKelvey 1999). Central-place foraging species leave from and return to a central location during foraging bouts (e.g., a nest; Orians and Pearson 1979). In this case, distance from the central location may influence an individual's use of a particular habitat type. Rosenberg and McKelvey (1999) found that traditional methods of assessing habitat selection were biased when applied to central-place foraging species, potentially leading to erroneous conclusions. Similarly, the degree to which a species is influenced by local disturbance may differ for central-place foragers. Allegheny woodrats are less sensitive to disturbance in the proximity of the rock outcrop, possibly because, as central place foragers, proportionally more of their movements occur closer to the outcrop. Whether other central-place foraging species exhibit a similar relationship with local disturbance is unknown, but foraging strategy of the species in question should be considered when applying conclusions herein to other species.

Although the literature suggests that local habitat disturbance has little effect on Allegheny woodrats, most of the information come from areas within distribution where population numbers are apparently stable. How those results relate to areas where populations are less stable, or how the results relate to other species, is unknown. LoGiudice (2006) suggested that multiple factors acting in concert have caused Allegheny woodrat declines in the northern portions of the distribution. For example, local habitat disturbance has little or no effect in areas where agricultural and urban development is limited. However, habitat disturbance where agricultural and urban development is high may result in an increase in raccoon abundance (Prange et al. 2005) which can facilitate transmission of raccoon roundworm (*Baylisascaris procyonis*; LoGiudice 2003). Although conducting research in more stable areas may be more conducive to obtaining adequate sample sizes from which to make inferences regarding declining populations, caution must be taken when applying those results to areas where other factors may be at work.

References

Balcom, B. J., and Yahner, R. H. 1996. Microhabitat and landscape characteristics associated with the threatened Allegheny woodrat. Conservation Biology 10:515–523.

Bommarito, M. P. 1999. Distribution of the Allegheny woodrat in the Daniel Boone National Forest: habitat characteristics and incidence of raccoon roundworm. M.S. Thesis, Eastern Kentucky University, Richmond, Kentucky.

Bowers, M. A., Welch, D. N., and Carr, T. G. 1990. Home range size adjustments by the eastern chipmunk, *Tamias striatus*, in response to natural and manipulated water availability. Canadian Journal of Zoology 68:2016–2020.

Castleberry, S. B., Ford, W. M., Wood, P. B., Castleberry, N. L., and Mengak, M. T. 2001. Movements of Allegheny woodrats in relation to timber harvesting. Journal of Wildlife Management 65:148–156.

Castleberry, N. L., Castleberry, S. B., Ford, W.M., Wood, P.B., and Mengak, M.T., and 1, M. T. 2002a. Allegheny woodrat (*Neotoma magister*) food habits in the central Appalachians. American Midland Naturalist 147:80–92.

Castleberry, S. B., Wood, P. B., Ford, W. M., Castleberry, N. L., and Mengak, M. T. 2002b. Summer microhabitat selection by foraging Allegheny woodrats (*Neotoma magister*) in a managed forest. American Midland Naturalist 147:93–101.

Castleberry, S. B., Mengak, M. T., and Ford, W. M. 2006. *Neotoma magister*. Mammalian Species 789:1–5.

Chamblin, H. D., Wood, P. B., and Edwards, J. W. 2004. Allegheny woodrat (*Neotoma magister*) use of rock drainage channels on reclaimed mines in southern West Virginia. American Midland Naturalist 151:346–354.

Clough, G. C. 1987. Relations of small mammals to forest management in northern Maine. Canadian Field-Naturalist 101:40–48.

Conditt, S. A., and Ribble, D. O. 1997. Social organization of *Neotoma micropus*, the southern plains woodrat. American Midland Naturalist 137:290–297.

Cranford, J. A. 1977. Home range and habitat utilization by *Neotoma fuscipes* as determined by radiotelemetry. Journal of Mammalogy 58:165–172.

Ford, W. M., Castleberry, S. B., Mengak, M. T., Rodrigue, J. L., Feller, D. J., and Russell, K. R. 2006. Persistence of Allegheny woodrats *Neotoma magister* across the mid-Atlantic Appalachian Highlands landscape, USA. Ecography 29:745–754.

Goertz, J. W. 1970. An ecological study of *Neotoma floridana* in Oklahoma. Journal of Mammalogy 51:94–104.

Hornsby, B. S., Ruiz, A. M., Castleberry, S. B., Castleberry, N. L., Wood, P. B., and Ford, W. M. 2005. Fall movements of Allegheny woodrats in harvested and intact stands in West Virginia. Northern Journal of Applied Forestry 22:281–284.

Ivanovich, E. S., Sumithran, S., and Frederick, R. B. 2001. Spatially explicit modeling of Allegheny woodrat habitat in the Daniel Boone National Forest, Kentucky. Proceedings of the Southeastern Association of Fish and Wildlife Agencies 55:364–374.

Jones, E. N. 1990. Effects of forage availability on home range and population density of *Microtus pennsylvanicus*. Journal of Mammalogy 71:382–389.

Kirkland, G. L., Jr. 1977. Responses of small mammals to clearcutting of northern Appalachian forests. Journal of Mammalogy 58:600–609.

Krupa, J. J., Workman, J. W., Lloyd, C. M., Bertram, L. R., Horrall, A. D., Dick, D. K., Brewer, K. S., Valentine, A. M., Shaw, C. P., Clemons, C. M., Prater, C. A., Clemons, J. E., Jr., Campbell, N. J., Arnold, S. B., Jones, N. J., and Clark, A. M. 2004. Distribution and habitat use of the Allegheny woodrat (*Neotoma magister*) in a mixed-mesophytic forest in southeastern Kentucky. Journal of the Kentucky Academy of Science 65:33–38.

Lacher, T. E., Jr., and Mares, M. A. 1996. Availability of resources and use of space in eastern chipmunks, *Tamias striatus*. Journal of Mammalogy 77:833–849.

LoGiudice, K. 2003. Trophically transmitted parasites and the conservation of small populations: raccoon roundworms and the imperiled Allegheny woodrat. Conservation Biology 17: 258–266.

LoGiudice, K. 2006. Toward a synthetic view of extinction: a history lesson from a North American rodent. Bioscience 56:687–693.

Lynch, M. F., Fesnock, A. L., and Van Vuren, D. 1994. Home range and social structure of the dusky-footed woodrat (*Neotoma fuscipes*). Northwestern Naturalist 75:73–75.

Mengak, M. T. 2002. Home range and movement of the Allegheny woodrat (*Neotoma magister*) in Virginia. Banisteria 19:3–8.

Monthey, R. W., and Soutiere, E. C. 1985. Responses of small mammals to forest harvesting in northern Maine. Canadian Field-Naturalist 99:13–18.

Myers, R. T. 1997. Microhabitat and ecology of the Allegheny woodrat in northcentral West Virginia. M.S. Thesis, West Virginia University, Morgantown, West Virginia.

Newcombe, C. L. 1930. An ecological study of the Allegheny cliff rat (*Neotoma pennsylvanica* Stone). Journal of Mammalogy 11:204–211.

Orians, G. H., and Pearson, N. E. 1979. On the theory of central place foraging. *In* Analysis of Ecosystems, eds. D. J. Horn, R. D. Mitchell and G. R. Stairs, pp.154– 177. Ohio State University Press, Columbus, Ohio.

Poole, E. L. 1940. A life history sketch of the Allegheny woodrat. Journal of Mammalogy 21: 249–270.

Prange, S., Gehrt, S. D., and Wiggers, E. P. 2005. Demographic factors contributing to high raccoon densities in urban landscapes. Journal of Wildlife Management 67:324–333.

Rosenberg, D. K., and McKelvey, K. S. 1999. Estimation of habitat selection for central-place foraging animals. Journal of Wildlife Management 63:1028–1038.

Sakai, H. F., and Noon, B. R. 1997. Between-habitat movement of dusky-footed woodrats and vulnerability to predation. Journal of Wildlife Management 61:343–350.

Taulman, J. F., Smith, K. G., and Thill, R. E. 1998. Demographic and behavioral responses of southern flying squirrels to experimental logging in Arkansas. Ecological Applications 8:1144–1155.

Topping, M. G., and Millar, J. S. 1996a. Home range size of bushy-tailed woodrats, *Neotoma cinerea*, in southwestern Alberta. Canadian Field-Naturalist 110:350–353.

Topping, M. G., and Millar, J. S. 1996b. Foraging movements of female bushy-tailed woodrats (*Neotoma cinerea*). Canadian Journal of Zoology 74:798–801.

Vaughan, T. A., and Schwartz, S. T. 1980. Behavioral ecology of an insular woodrat. Journal of Mammalogy 61:205–218.

Wood, P. B. 2001. Characteristics of Allegheny woodrat (*Neotoma magister*) habitat in the New River Gorge National River, West Virginia. Final Project report to the National Park Service, New River Gorge National River, Glen Jean, West Virginia.

Yahner, R. H. 1986. Microhabitat use by small mammals in even-aged forest stands. American Midland Naturalist 115:174–180.

Chapter 5
Den Use Behavior of Allegheny Woodrats Inhabiting Rock Outcrops in Pennsylvania

John D. Peles and Janet Wright

Introduction

Since the decline of the Allegheny woodrat (*Neotoma magister*) was first noticed in the late 1970s, most information concerning the ecology of this species has been acquired from studies focusing on its association with rocky habitats (Newcombe 1930, Poole 1940, Hall 1985, Wright and Butchkoski 2005, Castleberry et al. 2006). This habitat specialization has been implicated in microhabitat selection during foraging (Castleberry et al. 2001), home range movements (Castleberry et al. 2002, Hornsby et al. 2005), and persistence of woodrat populations (Balcom and Yahner 1996, Ford et al. 2006).

Less attention has been given to understanding how behavioral characteristics of *N. magister* may render this species vulnerable to decline and present a challenge for conservation and management. The importance of integrating behavioral ecology with conservation biology has received increased attention in recent years (Caro 1998, Morris 2001) and has been applied to conservation questions involving small mammals such as the Stephens' kangaroo rat (*Dipodomys stephensi*; Brock and Kelt 2004) and the riparian woodrat (*N. fuscipes riparia*; Gerber et al. 2003).

There are indications that behavioral characteristics are relevant to the conservation ecology of *N. magister* as well. For example, Castleberry et al. (2001) showed that Allegheny woodrats willingly forage in timber clearcuts adjoining the rock patches they inhabit. While this may have been a benign or even advantageous behavior in former times when occasional fires or blowdowns created temporary clearings, the same tendency may now expose foraging woodrats to increased predation when large expanses of the forest matrix are routinely removed (Castleberry Chapter 4). LoGiudice (2001) demonstrated that *N. magister* harvest dried, seed-loaded scats such as those of raccoons, retrieving them intact to a food cache rather than consuming them when found. While an appropriate behavior in terms of nutrition, this tendency maximizes exposure of woodrats using the cache to the

J.D. Peles
Pennsylvania State University-Greater Allegheny, McKeesport, PA 15132

J.D. Peles, J. Wright (eds.), *The Allegheny Woodrat*, DOI: 10.1007/978-0-387-36051-5_5,

lethal roundworm *Baylisascaris procyonis* harbored in raccoon feces (LoGiudice 2001).

In this chapter, we examine how a particular set of behaviors—those that determine the use and spacing of dens in rock habitat—influences Allegheny woodrat ecology. Individual woodrats construct nests of plant fibers in caves or large crevices within rocks (Newcombe 1930, Wright and Butchkoski 2005). Nests are also likely to be constructed beneath the surface of rock outcrops or other rocky habitats that are not accessible to observers. Similar habits of nest construction within rocky habitats are exhibited by the bushy-tailed woodrat (*N. cinerea*; Escherich 1981, Hickling 1987), and several studies point to availability of suitable den sites as a limiting factor for local population size of this species (Escherich 1981, Hickling 1987, Moses and Millar 1992, Topping and Millar 1996a,b).

To complicate the picture, Moses and Millar (1992) have suggested that agonistic behavior by *N. cinerea* influences population size at rock outcrops because not all potential den sites are used. Similar observations for Allegheny woodrats indicate that intraspecific behaviors or individual preferences could be influencing den use. For example, Allegheny woodrats are known to be aggressive toward nonkin (Kinsey 1976, 1977) and this might limit proximity of simultaneously occupied den sites. Wright (1998) observed that woodrats change den sites frequently and suggested that they preferentially use certain den sites when available. Thus, behavioral characteristics of woodrats could potentially further limit the size of a local population within an area that already has a finite number of den sites.

Knowledge of specific ecological requirements of *N. magister* is critical to the development of a long-term strategy for the conservation and management of this species. In this chapter, we describe results from an intensive four-month radiotelemetry study of den use behavior by adult Allegheny woodrats inhabiting rock outcrops in southwestern and central Pennsylvania. Specifically, the purpose of the investigation was to determine: (1) frequency of den relocations and proportion of time spent at den sites by individual woodrats; (2) frequency of occupation for individual dens; and (3) spatial distribution of simultaneously occupied dens during a four-month period. Results of this investigation, along with findings from our long-term studies of den use in woodrats, are used to develop a conceptual model of the influence of den selection and den use behaviors of woodrats on the dynamics of local populations and metapopulations. We then apply our findings to the conservation and management of the Allegheny woodrat as well as other declining species.

Short-Term Study of Den Use

Study Sites

Woodrats were studied over a period of four months at three sites—two (Strangford Cave and Strangford Mountain) in southwestern Pennsylvania and one (Cove Mountain) in the central portion of the state. Strangford Cave (Indiana County, 40°25′ N,

79°12′ W) is an abandoned quarry along the Conemaugh River and is characterized by rock highwall with numerous large openings ("caves") and crevices with a series of sandstone boulders along the base of the highwall. Strangford Mountain (Indiana County, 40°25′ N, 79°10′ W) and Cove Mountain (Perry County, 40°21′ N, 76°56′ W) consist of sandstone boulder outcrops along ridgetops.

Methods

Den fidelity and den selection behavior of woodrats were monitored by radiotelemetry at each site during an early summer period (P-I) and a late summer/early fall period (P-II) in 2002. At Strangford Cave and Strangford Mountain, P-I covered 21 June–26 July, and P-II was 19 August–11 September. At Cove Mountain, P-I covered 17 June–24 July, and P-II was 18 August–12 September. The end of P-I and the beginning of P-II were separated by a hiatus of three or four weeks at each site. Following the end of P-II, woodrats at Cove Mountain were monitored 1–2 times per week until 6 November 2002.

Live-trapping was conducted prior to P-I, between monitoring periods, and following P-II. During each live-trapping period, Tomahawk live traps baited with sliced apples were used, and traps were set for two nights at each site ($n = 40$ traps per night). This trapping protocol was based on experience demonstrating two nights of live-trapping was sufficient to capture virtually all woodrats present. For example, during studies between 1994 and 2006 at these sites, we conducted 16 sessions with multiple-night trapping (three or four nights within a week). Of 97 captures in these sessions, only two woodrats were first caught after the second night of trapping.

Upon first capture, body mass, sex, and reproductive condition were determined, and each individual was given an identifying mark with an ear tattoo. Woodrats used for radiotelemetry were fitted with a 3.5 g temperature-sensitive radiotransmitter (PD-2CT; Holohil Systems, Ltd., Carp, Ontario) and released at the point of capture. All adults (>225 g; Mengak 2002) captured prior to P-I and between monitoring periods were fitted with radiotransmitters. Thus, the total number and sex ratio of woodrats monitored at each location and during each period varied based on the number of adult woodrats present. For analysis, individuals present for the entire study period (P-I through P-II) were classified as residents and those present during only P-I or P-II were classified as nonresidents.

Radiotelemetry was used to track woodrats to their den locations by homing (White and Garrott 1990) during daylight hours, and geographic coordinates, determined by GPS, were recorded for each location. Study sites were visited three times per week during each of the monitoring periods with no more than two days between visits. The location of all radio-collared individuals was recorded for each visit to a study site. Because the location of individuals was determined during daylight hours when woodrats are not active, these locations were presumed to represent den sites. Dens located within the main rock outcrop were classified as central dens and those associated with isolated rock formations separated from the main rock outcrop by >25 m were classified as peripheral dens.

Whenever a woodrat was tracked to a new location, the date of movement was estimated as the midpoint between the last known date of occupancy in the previous den and the first date that it was tracked to the new den. Maximum percentage of time spent at any given den was determined for each woodrat. This value was calculated by expressing the greatest number of days for which an animal was estimated to have occupied a den without moving as a percentage of the total days the animal was known to be present at the rock outcrop.

Multi-Year Studies of Den Use

In order to better understand the use of den sites and its associated behavior, we also examined data from several long-term studies of den use conducted by one of us (J. Wright) at other locations in central Pennsylvania. Study sites included Stoney Mountain (40°25′ N, 76°48′ W) in Dauphin County, Waggoners Gap (40°16′ N, 77°16′ W) in Cumberland/Perry Counties, and Bowers Mountain (40°16′ N, 77°30′ W) in Perry County. Each site is a sandstone boulder outcrop located along a ridgetop.

Radiotelemetry was conducted over extended periods of time and was used to determine the location of woodrat dens during the following time periods: Stoney Mountain: May–September 1995 (n = 5 woodrats), April–October 1996 (n = 4); Waggoners Gap: July 1996–June 1998 (n = 1); May–December 1997 (n = 6); Bowers Mountain: July–November 2002 (n = 7), June–November 2003 (n = 4), October 2004–March 2005 (n = 4). The frequency at which den locations were determined varied among study sites (three times per week to once per month), and not all woodrats were present for the entire time period. Because of differences in the frequency of tracking and the length of study periods for the multi-year studies, statistical analyses were not performed for these data. However, results from these studies are presented, where appropriate, to provide additional insight into the den use behavior of woodrats.

For one aspect of the analysis—the influence of season on the tendency of woodrats to change dens—the den location data from the short-term (Cove Mountain) and multi-year studies in central Pennsylvania were combined. Each change of den from the previous tracking session for an individual was categorized by month. An index of den-changing frequency for each month of the year was then developed by expressing the total number of den changes during each month as a fraction of the total number of tracking observations for that month.

Results and Discussion

Live-Trapping

Results of live-trapping at Strangford Cave and Strangford Mountain (Fig. 5.1) reflect the presence of several juveniles prior to P-I (body mass $\overline{X}$ = 174 g;

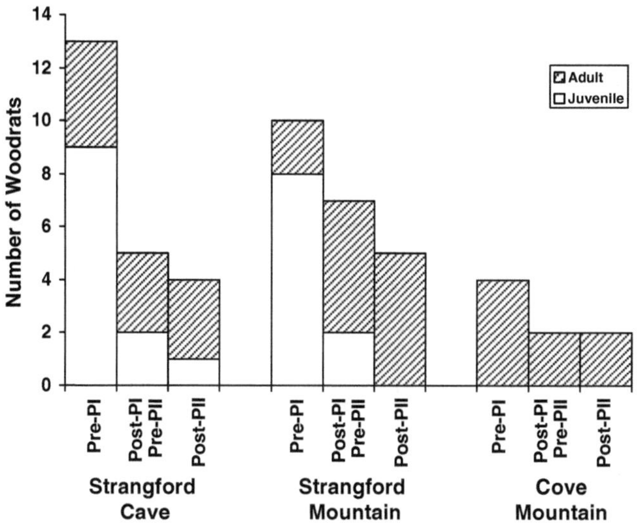

Fig. 5.1 Number of adult and juvenile Allegheny woodrats (*Neotoma magister*) captured at three sites (Strangford Cave, Strangford Mountain, and Cove Mountain) in Pennsylvania during three live-trapping sessions in 2002. Live-trapping sessions (Pre P-I = mid-June; Post P-I/Pre P-II = mid August; Post P-II = mid-September) were conducted before and after two radiotelemetry periods (P-I and P-II) at each site. Number of woodrats captured during each live-trapping session is based on 120 trap nights (3 nights per site × 40 traps per night)

range = 105–210 g). Two juveniles (200 g each) remained at these sites at the end of P-I and only one ($\overline{X} = 220$ g) remained at the end of P-II. No juvenile woodrats were trapped at Cove Mountain during the study.

Conclusions regarding den use and selection behavior of radio-collared individuals are valid only if these behaviors are not influenced by the presence of other adult woodrats whose movements were not monitored. No additional adults were captured at Strangford Cave and Cove Mountain at the end of P-I or P-II, and none were captured at Strangford Mountain at the end of P-II (Fig. 5.1). However, three previously unknown adults were present at Strangford Mountain following P-I. Although it is unknown when these individuals arrived at the site, any influence that they might have had on movements appears to be minimal given that the radio-collared adults at the site did not change dens during P-I.

Den Locations

Den locations ($n = 286$) were determined for 12 adult woodrats (5 males, 7 females) during the investigation, and 7 (3 males, 4 females) individuals were present during both P-I and P-II. Twenty-seven different dens were used by the 12 woodrats monitored at Strangford Cave, Strangford Mountain, and Cove Mountain (Table 5.1). Approximately one-fourth (25–29%) of dens at each site were classified as

Table 5.1 Summary of dens used by Allegheny woodrats (*N. magister*) at study sites in Indiana County (Strangford Cave, Strangford Mountain) and Perry County (Cove Mountain), Pennsylvania during 2002. Number of reused dens refers to dens used by more than one woodrat, but at different times

Site	Total known dens	Central dens	Peripheral dens	Maximum percent of known dens occupied (%)	Number of reused dens
Strangford Cave	12	9	3	33	2
Strangford Mountain	8	6	2	63	2
Cove Mountain	7	5	2	43	1
All Sites	27	20	7	44	5

peripheral. The proportion of total observations occurring at peripheral dens did not differ significantly between time periods ($\chi^2 = 2.84, p > 0.05$) or between residents and nonresidents ($\chi^2 = 3.74, p > 0.05$). The percentage of peripheral dens at Stoney Mountain (22% of 9 den sites) was similar to the three short-term sites, whereas peripheral dens represented a much higher percentage of den sites at Bowers Mountain (59% of 17 dens) and Waggoners Gap (69% of 13 dens).

Male woodrats (49%) were located for a significantly ($\chi^2 = 36.15, p < 0.001$) greater proportion of the time at peripheral dens compared to females (23%). Occupancy of central dens by males was limited to those that were a minimum of 40 m from the nearest female-occupied den. We also observed two cases where a female woodrat apparently displaced a male from a central den. Female territoriality has been documented in the southern plains woodrat (*N. cinerea*; Conditt and Ribble 1997) and eastern woodrat (*N. floridana*; Goertz 1970). Although territoriality has not been studied in *N. magister*, territorial behavior by females is suggested by the fact that this sex occupies smaller home ranges than males (Castleberry et al. 2002, Castleberry Chapter 4) and are aggressive toward nonkin (Kinsey 1976, 1977). Thus, our results suggest the restriction of males to peripheral habitats by female territoriality.

There was little visual evidence (i.e., presence of scat or caches) of woodrats in the small, isolated rock formations that contained peripheral dens, and these sites are typically not the focus of live-trapping efforts. However, peripheral den sites represent landscape features that should not be overlooked in the development of management plans for this species. At the very least, the availability of peripheral dens around one or more large rock outcrops would maximize the size of a local woodrat population. This is especially relevant for areas such as Waggoners Gap and Stoney Mountain where a high percentage of dens are peripheral. Furthermore, if males are restricted to peripheral den sites by female territoriality, these areas would be necessary for males to remain near rock outcrops and have access to females during periods of reproduction. Peripheral den sites also represent landscape elements that could increase connectivity and facilitate the movement of individuals between rock outcrops in a local population or between two local populations in a metapopulation (Fahrig and Merriam 1985).

Quantity and Quality of Den Sites

Results indicate that the number of individuals inhabiting a rock outcrop and its associated peripheral areas was not limited by the number of den sites. Of the total number of dens documented at Strangford Cave, Strangford Mountain, and Cove Mountain, the maximum percent of dens occupied during any given time was 63% (Table 5.1). Similarly, maximum percent occupancy at the multi-year study sites ranged from 24 to 56%.

Species of North American *Neotoma* exhibit both a den-centered ecology and habitat specificity that creates the potential for the availability of den sites to limit the density of breeding adults within local populations (Topping and Millar 1996a, Dial 1988). The influence of number of den sites on local population size has been documented in several studies of *N. cinerea* inhabiting rock outcrops in western Canada (Escherich 1981, Hickling 1987, Moses and Millar 1992, Topping and Millar 1996b). Support for this notion also is provided by results of studies involving other species of *Neotoma*, such as the eastern woodrat (*N. floridana floridana*; HaySmith 1995) and white-throated woodrat (*N. albigula*; Newton 1990) that have very different habitat requirements from *N. magister*. Thus, the habitat-specific nature of *N. magister* (Newcombe 1930, Poole 1940, Hall 1985, Castleberry et al. 2006) makes it likely that the maximum size of local populations of this species also is limited by the availability of den sites. However, total quantity of den sites alone is probably no longer a limiting factor in regions such as southwestern and central Pennsylvania where woodrat populations have declined (Hall 1985; Wright and Butchkoski 2005, Castleberry et al. 2006).

At least one den at each study site was used by more than one woodrat during the investigation (Table 5.1). All five of the dens occupied by more than one woodrat were central dens and four were occupied by different woodrats in successive tracking sessions. The high level of use for some specific dens suggests the existence of preferred den sites. McGowan (1993) also observed occupancy of certain dens in succession by more than one individual in a study of translocated Allegheny woodrats in New York. Similar observations have been made in studies of *N. f. floridana* (HaySmith 1995) and *N. micropus* (Merkelz and Kerr 2002).

The existence of preferred den sites in this and other investigations suggests differences in den quality (McGowan 1993, HaySmith 1995, Merkelz and Kerr 2002). Woodrats use dens for a number of reasons including shelter, protection from predation, rearing of young, and storing of food caches (Hickling 1987, Kelly 1989, Newton 1990). Variation in the suitability of individual den sites for these and other functions likely results in differential preference for certain dens. Although the quality of a den site might be enhanced by the presence of a nest or food cache left by the previous occupant (Poole 1940), there are a variety of other physical (e.g., microclimate) and biological (e.g., access to foraging sites) factors that could affect the suitability of a den for meeting specific needs. The availability of high-quality den sites would be especially important for female woodrats occupying centrally located dens on the main rock outcrop during periods of reproduction. Dens of poor quality also could limit the amount of time that an individual may occupy the den or may even lead to increased mortality. Mortality of *N. lepida* has been shown to be correlated with decreasing den quality (Brown et al. 1972).

Den Fidelity and Behavior

Although the mean maximum percentage of time spent by an individual at any given den was relatively high (68%), nearly all woodrats at Strangford Cave, Strangford Mountain, and Cove Mountain changed dens at least once (Table 5.2). In fact, only two individuals remained in the same den for the entire four-month period, and the mean number of dens used per woodrat was greater than 1.0 at each of the study sites. Similarly, in the multi-year studies, 50% of all individuals moved at least once, and the mean number of dens used per woodrat was greater than 1.0 at all sites.

At the three short-term sites, number of den relocations did not differ significantly between males and females ($t = 0.54$, $p > 0.05$) or between P-I and P-II ($t = 0.43$, $p > 0.05$). These results are in contrast with those from a similar study of den relocation in *N. cinerea* (Holmes 1995). In that investigation, Holmes (1995) used radiotelemetry to track den locations of 11 adult *N. cinerea* from July through November. All females except one each maintained a single den throughout the study period. In contrast, all males changed dens during the study and averaged nearly three den relocations per individual (Holmes 1995).

Although number of relocations was not significantly different for residents compared to nonresidents ($t = 1.92$, $p > 0.05$), nonresidents ($\overline{X} = 2.8 \pm 1.6$ SD) changed dens more than twice as often as residents (1.1 ± 1.3). Mean maximum percent time spent at any given den differed significantly ($t = 4.11$, $p = 0.003$) between nonresidents (47.1 ± 8.2) and residents (83.3 ± 20.9). It is not known whether individuals that disappeared by the end of P-I represented over-wintering adults or those that immigrated to the sites during the early spring. However, differences in den fidelity and movements between residents and nonresidents suggest that at

Table 5.2 Summary of den use behavior by Allegheny woodrats (*N. magister*) at study sites in Indiana County (Strangford Cave, $n = 4$; Strangford Mountain, $n = 5$) and Perry County (Cove Mountain, $n = 3$), Pennsylvania during 2002

Site	Mean number of different dens used	Mean number of relocations	Mean maximum percent time at a den (%)	Mean distance moved (m)	Percent using multiple dens (%)
Strangford Cave	3.0	2.0	72	61.6	50
Strangford Mt.	1.8	1.2	70	24.5	80
Cove Mountain	2.3	2.7	61	50.0	100
All Sites	2.3	1.8	68	47.3	83

least some adults could be forced to leave the rock outcrop or experience mortality because of difficulty in becoming established at a den site. Similarly, adults entering the rock outcrop prior to P-II exhibited much greater movement among dens than those present during the entire study period. These increased movements by some individuals could reflect a lack of unoccupied quality den sites and/or exclusion from quality sites on the main rock outcrop as a result of agonistic behavior.

Female aggression toward nonkin individuals is exhibited by *N. magister* (Poole 1940, Kinsey 1976, 1977, Castleberry et al. 2002) and other species of *Neotoma* (Dial 1988, Moses and Millar 1992, Conditt and Ribble 1997). In fact, *N. magister* has been shown to form dominance hierarchies in confined populations and both males and females exhibit agonistic behavior toward all conspecifics (Kinsey 1976, 1977). Thus, agonistic behavior could account for the apparent difficulty of nonresidents in becoming established within an occupied den site. This behavior would also contribute to the low rate of juvenile persistence in rock outcrops that we have observed in this and other investigations (Fig. 5.1; J. Peles and J. Wright pers. obs.).

The mean nearest-neighbor distance between simultaneously occupied dens at the three study sites was 89 m, and nearest-neighbor distances for some dens were greater than 150 m (Fig. 5.2). Mean nearest-neighbor distance between dens simultaneously occupied by two females ($\overline{X} = 81$ m) did not differ significantly ($t = 0.12, p > 0.05$) from those occupied by a male and female (78 m). Because there were relatively few pairs of simultaneously occupied dens for males ($\overline{X} = 66$ m), these were not compared statistically to male/female or female/female pairs. In contrast, mean nearest-neighbor distance between dens occupied by a resident and

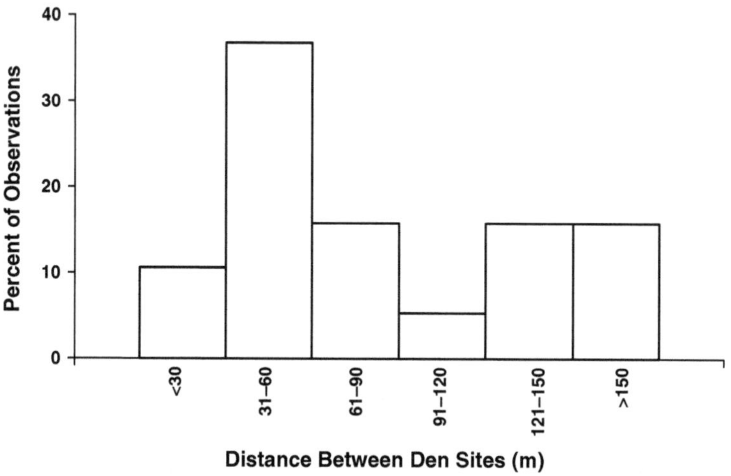

Fig. 5.2 Summary of nearest-neighbor distances between pairs of dens simultaneously occupied by Allegheny woodrats (*Neotoma magister*) at three locations (Strangford Cave, Strangford Mountain, and Cove Mountain) in Pennsylvania during June through September 2002. Results are summarized as the percent of all observations represented within each of six distance categories

nonresident ($\overline{X} = 61$ m) differed significantly ($t = 2.96$, $p = 0.02$) from the distance between two residents ($\overline{X} = 143$ m). Mean distance between nonresident dens ($\overline{X} = 41$ m) was significantly ($t = 2.89$, $p = 0.03$) lower compared to resident dens but did not differ significantly ($t = 0.90$, $p > 0.5$) from the distance between nonresident/resident dens.

In the context of agonistic behavior, it might be expected that there exists some minimum distance tolerated between two simultaneously occupied dens. Evidence to support this expectation is provided by the fact that the mean nearest-neighbor distance between dens occupied by residents was more than 2.5 times greater than the distance between resident/nonresident dens. Dens occupied by two residents represent sites that the occupants are less likely to vacate (i.e., residents change dens less often) whereas resident/nonresident pairs represent a situation where one of the dens is likely to be vacated (i.e., nonresidents change dens more frequently). In fact, occupancy of dens separated by less than 30 m accounted for only 11% of our observations (Fig. 5.2) and, in all cases, one of the dens was occupied by a resident and the other by a nonresident that eventually left the den. In contrast, the greater distance between resident dens apparently minimizes the probability of agonistic interactions.

Our findings suggest that behavioral interactions could have a significant influence on the population ecology of Allegheny woodrats. Specifically, agonistic behaviors among woodrats could limit the long-term residence of breeding adults at a rock outcrop by influencing the use of quality of den sites. Similar conclusions have been drawn from an investigation of *N. cinerea* (Moses and Millar 1992). More recently, Gerber et al. (2003) has determined that the spatial arrangement of artificial den sites—reflecting behavioral interactions among individuals—was an important factor in the success of translocation experiments for *N. fuscipes riparia*.

Despite the fact that a relatively high degree of den fidelity was exhibited by residents compared to nonresidents, even most of these individuals changed dens at least once. In three cases, an individual changed dens, and eventually moved back to a den that it had previously occupied. Similar behavior also has been documented for other species of woodrats (HaySmith 1995, Conditt and Ribble 1997, Merkelz and Kerr 2002). Although the reason for this behavior is uncertain, adult woodrats exhibit an apparent need to occupy more than one den over a relatively short period of time. It is even possible that woodrats defend more than one den site at a time, something we cannot determine by conducting radiotelemetry during daylight hours. This behavior might further limit the number of breeding adults that could occupy a rock outcrop at any given time.

The tendency to shift dens appeared to have a distinct seasonal component. Nearly all (92%) of the den shifts that we observed in the multi-year studies and in the extended monitoring period at Cove Mountain took place in the late spring to summer months (May–August). In contrast, only 59% of our total observations took place during these months (Fig. 5.3). No woodrats shifted den locations between mid September and late January. Although our data did not meet assumptions of independence necessary for statistical analysis, the striking seasonal pattern we observed suggests that this is an important aspect of woodrat behavior that

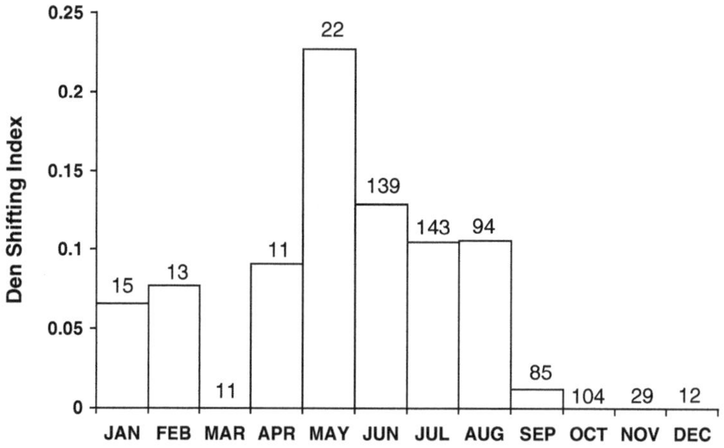

Fig. 5.3 Seasonal pattern of den shifting for Allegheny woodrats (*Neotoma magister*) at four sites in central Pennsylvania. Den shifting (*Y*-axis) is the proportion of total radiotracking observations for each month in which a woodrat changed den location since the previous observation. Number of radiotracking points for each month is shown above bar

deserves further study. In Pennsylvania and West Virginia, Allegheny woodrats begin storing winter food caches in late summer (Castleberry 2000). During winter they do not hibernate but remain active, consuming their stored food cache, and reproduction typically commences in early spring (Poole 1940). A litter is nursed for about a month, after which juveniles may remain with their mother for sometime but do their own foraging (Poole 1940). As far as is known, young do not remain at the maternal den through their first winter, but probably must find and provision their own winter den site. The summer den shifting we observed in adult woodrats may represent exploration or competition for position to establish both a secure winter refuge, and for females, an appropriate site for rearing the next spring's litter. After mid-September, a shift to a new den would leave little time to establish a winter food cache, and the safest strategy may be to hold on to the current den through winter.

Influence of Behavior on Population Dynamics

A Proposed Model

Although historical data are not available from all study sites, as many as 50 woodrats were captured during three-night trapping periods at Strangford Cave in the mid-1970s (A. Linzey pers. comm.). Thus, the number of woodrats (3–13) we documented at any given time at our study sites was likely much less than existed prior to the decline of this species in Pennsylvania. Numbers that are far below the expected capacity of a rock outcrop make it difficult to draw definitive conclusions about the influence of behavior on ecological processes. However, our findings, considered with known aspects of woodrat behavior, strongly suggest that den

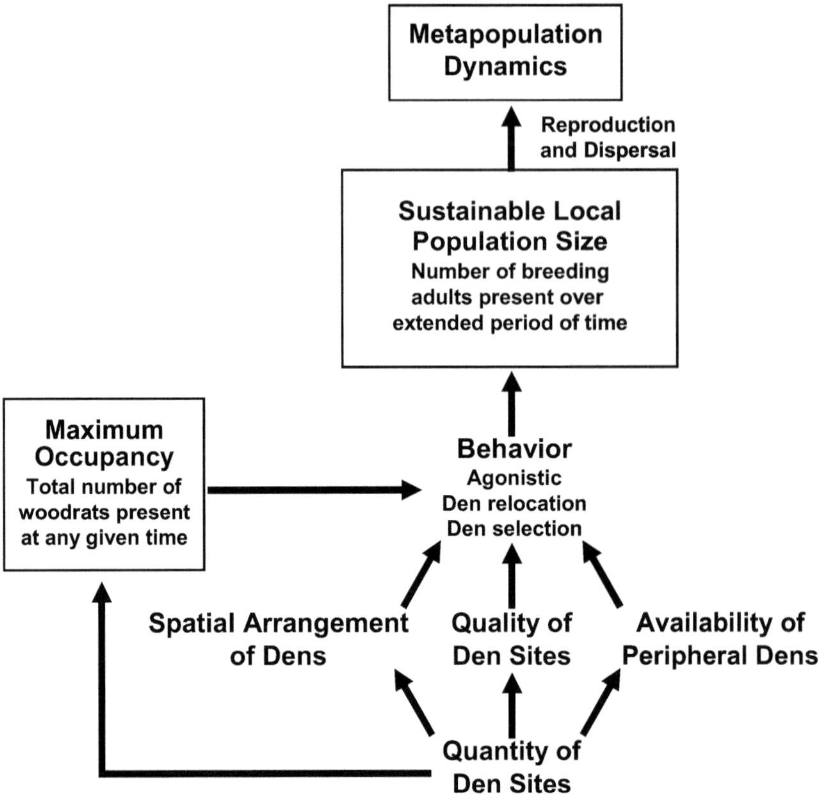

Fig. 5.4 Conceptual model of the possible influence of den characteristics and behaviors on the size of local and regional populations of the Allegheny woodrat (*Neotoma magister*)

use and den selection could have a significant influence on population dynamics of Allegheny woodrats.

We propose an empirically testable model for understanding the influence of behavior on local population size that distinguishes between maximum occupancy and sustainable population size (Fig. 5.4). Maximum occupancy reflects the potential number of woodrats present at a rock outcrop at any given time and includes resident adults, transients, and juveniles. Because Allegheny woodrats are habitat-specific (Newcombe 1930, Poole 1940, Hall 1985, Wright and Butchkoski 2005, Castleberry et al. 2006), the maximum number of individuals present is expected to depend on the quantity of den sites.

In contrast to maximum occupancy, sustainable local population size reflects the number of breeding adults residing at a rock outcrop over an extended period of time. Although we clearly did not observe maximum occupancy at any study site, support for the notion of a dichotomy between maximum number and sustainable population size is provided by the observation that nearly all juveniles had disappeared from the sites by mid-August (Fig. 5.1). In addition, results indicated that some adults (i.e., nonresidents) had difficulty in becoming established at any given den site for an extended period of time and at least some disappeared from the rock outcrop.

Whereas quantity of dens determines maximum occupancy, three characteristics of dens and behaviors associated with their use would limit the sustainable size of a population at a rock outcrop (Fig. 5.4):

(1) *Quality of den sites*—Differences in the quality of dens might influence behavior in several ways. For example, woodrats would be expected to select dens preferentially based on quality and to exhibit aggressive defensive behavior at these sites. In addition, the frequency of den relocations by an individual woodrat would be increased if poor quality dens are unable to support a woodrat for an extended period of time or if den quality decreases over time. Therefore, even if many dens are available, various aspects of quality and behaviors related to the use of these dens would reduce the number of woodrats that could be supported at a rock outcrop.
(2) *Spatial arrangement of dens*—The observation that the minimum distance between den sites of resident woodrats averaged 103 m likely reflects agonistic behavior of woodrats toward conspecifics (Kinsey 1976, 1977). If there exists a minimum distance between den sites, then local population size would be maximized by a spatial arrangement that maximizes distances between sites. Thus, even in a rock outcrop with many high quality den sites, sustainable population size would be limited by agonistic behavior.
(3) *Availability of peripheral dens*—In addition to the need for a minimum distance, results suggest that male woodrats may be restricted to peripheral den sites that are not located on the main rock outcrop. In this scenario, sustainable population size is limited by the availability of peripheral sites that allow male woodrats to establish permanent residence and permit access to females during the breeding season.

The concept of a metapopulation reflects the existence of local populations in a given area that are connected by dispersal (Hanski 1996). This concept fits well with our current knowledge of the Allegheny woodrat in Pennsylvania and elsewhere. In this case, behavioral processes that influence local populations would also influence the dynamics of the metapopulation (Batzli et al. 1999). For example, increased sustainable population size at individual outcrops would increase the probability that other outcrops in the metapopulation would be colonized when local extinctions occur (Fig. 5.4).

Testing the Model

The small size of *N. magister* populations in southwestern and central Pennsylvania make it difficult to distinguish normal population and behavioral processes from stochastic events (Hanski 1996). Consequently, empirical tests of predictions concerning the influence of den selection and den use behaviors on the size of local populations of *N. magister* will require data collected from healthy populations where maximum occupancy of rock outcrops might be expected at certain times of the year (e.g., late spring). Ideally, these investigations would be conducted in conjunction with studies from areas where the species has declined to permit comparison of woodrat behavior between rock outcrops that differ with respect to population density.

Relatively little is known concerning changes in woodrat population size and demographic composition during a single year or over the course of several years (Wood Chapter 3). The seasonal changes in den use we observed, with little or no moving to new dens in fall or winter, suggests that there may be a seasonal bottleneck in high-quality dens that can affect over-winter survival and spring reproduction. Therefore, live-trapping data collected over a time span including the early part of the breeding season through the early fall during several years are needed. These data would permit us to determine if there is consistency in woodrat population dynamics at a rock outcrop over time and if our dichotomy between maximum occupancy and sustainable population size is valid. Furthermore, information concerning demographic composition would permit us to determine if there exists a consistent ratio of breeding adults from one year to the next.

The patterns we observed, and on which our model is based, need to be confirmed with radiotelemetry studies in areas where woodrats have not experienced recent decline. Within these populations, our model predicts: (1) the existence of a minimum distance between den sites; (2) the existence of preferred den sites that are used disproportionately; (3) differences in den use behavior between sexes including predominant use of peripheral dens by males; (4) differences in the frequency of den relocation between residents and nonresidents; (5) periodic changing of dens by all woodrats; and (6) seasonal variation in den shifting. It is important that these predictions be addressed in conjunction with data concerning population dynamics and demographic characteristics.

Although not specifically required to test our model, information concerning various aspects of woodrat ecology and genetics would greatly enhance our understanding of the influence of behavior on population dynamics. For example, information concerning genetic relatedness among individuals, considered in conjunction with distances between individual den sites, would provide insight into the influence of social relationships on den selection. Data concerning the fate of juveniles and adults that disappear from the outcrop would be important to determine if these individuals disperse, suggesting that they are forced to leave the outcrop, or suffer mortality. Lastly, information concerning the relationship between survivorship and various aspects of den quality would provide support for the significance of preferred den sites.

Conservation and Management of Declining Species

In this chapter, we have proposed a model of how behaviors related to the selection and use of dens by Allegheny woodrats could influence the dynamics of populations at individual rock outcrops as well as the dynamics of metapopulations.

Consequently, the behavior of *N. magister* deserves consideration in the context of conservation and management of this species. We further suggest that behavior may be an important factor for many other species in decline. For example, our model is clearly applicable to other species of *Neotoma*. This is particularly relevant given the fact that regional populations or subspecies of several species of North American *Neotoma* have experienced declines in the past 50 years (Feldhamer and Poole Chapter 11).

Like *N. magister*, most species of *Neotoma* are habitat specialists and exhibit den-centered ecology within selected habitats (Dial 1988, Topping and Millar 1996a). Habitat specificity is considered a characteristic that is correlated with susceptibility to extinction processes (Terborgh 1974) and the need for appropriate den sites within selected habitats further limits the size of a local population (Dial 1988, Topping and Millar 1996a). Behaviors such as agonistic defense of den sites, selection for den quality, and frequent den relocation would further limit the number of individuals inhabiting a local population. In this regard, behavior further limits the use of an already limiting resource and makes regional populations more susceptible to decline.

Behaviors that could make a species susceptible to decline also could complicate conservation efforts for that species. For example, re-establishment of woodrats at sites that have been subjected to local extinction would require consideration of the spatial arrangement of den sites to minimize the likelihood of agonistic interactions among individuals. In their evaluation of the potential for the translocation of a population of the endangered riparian woodrat (*N. fuscipes riparia*), Gerber et al. (2003) concluded that the distribution of potential dens and configuration of habitat structures was an important consideration based on the behavior of the species. Closer study of the determinants of a high-quality den site is needed for an informed choice of reintroduction sites, and especially to open the possibility of den-site enhancement. In the case of *N. magister* in southwestern and central Pennsylvania, this consideration is further complicated by the fact that many outcrops are linear in nature. Efforts to conserve declining local populations or reintroduce woodrats to extirpated populations should also consider that a main rock outcrop alone may not be sufficient but that associated peripheral sites for males might be necessary to support a breeding population (Serfass Chapter 10).

Our studies of den selection and den use in *N. magister* and our subsequent model of the influence of behavior on population dynamics (Fig. 5.4) address behaviors that apply specifically to *N. magister* and, in more general terms, to other species of *Neotoma*. However, findings from our study of *N. magister* suggest general principles that are applicable to the conservation and management of many declining species. For example, behaviors related to habitat use may limit local population size and make these populations more susceptible to stochastic processes and local/regional extinctions. This is especially true for species that are highly habitat-specific (Terborgh 1974) and exhibit a patchy distribution throughout the landscape. Furthermore, many behavioral processes (i.e., sociality, habitat selection, habitat use) may represent important considerations for management or reintroduction efforts. Similar conclusions have been drawn in a recent study of the Stephens' kangaroo rat (Brock and Kelt 2004). Thus, the opportunity to study both declining and healthy populations of *N. magister* will provide us further insight into the relationships between behavior and conservation of declining species.

Acknowledgments Funding was provided by a grant from the Pennsylvania Department of Conservation and Natural Resources' Wild Resource Conservation Fund. Dickinson College students, especially Laura Pell, Chris Magel, and Lindsay Wieland, contributed to data collection and analyses. Carolyn Mahan provided helpful comments on an earlier draft of the manuscript.

References

Balcom, B. J., and Yahner, R. H. 1996. Microhabitat and landscape characteristics associated with the threatened Allegheny woodrat. Conservation Biology 10:515–525.

Batzli, G. O., Harper, S. J., Lin, Y. K., and Desy, E. A. 1999. Experimental analyses of population dynamics: scaling up to the landscape. *In* Landscape Ecology of Small Mammals, eds. G. W. Barrett and J. D. Peles, pp. 107–127. Springer-Verlag, New York.

Brock, R. E., and Kelt, D. A. 2004. Conservation and social structure of Stephens' kangaroo rat: implications from burrow-use behavior. Journal of Mammalogy 85:51–57.

Brown, H. H., Lieberman, G. A., and Dengler, W. F. 1972. Woodrats and cholla: dependence of a small mammal population on the density of cacti. Ecology 53:310–313.

Caro, T. (ed.). 1998. Behavioral Ecology and Conservation Biology. Oxford University Press, New York.

Castleberry, N. L. 2000. Food habits of the Allegheny woodrat (*Neotoma magister*). M.S. Thesis, West Virginia University, Morgantown, West Virginia.

Castleberry, S. B., Ford, W. M., Wood, P. B., Castleberry, N. K., and Mengak, M. T. 2001. Movements of Allegheny woodrats in relation to timber harvesting. Journal of Wildlife Management 65:148–156.

Castleberry, S. B., Wood, P. B., Ford, W. M., Castleberry, N. L., and Mengak, M. T. 2002. Summer microhabitat selection by foraging Allegheny woodrats (*Neotoma magister*) in a managed forest. American Midland Naturalist 147:93–101.

Castleberry, S. B., Mengak, M. T., and Ford, W. M. 2006. *Neotoma magister*. Mammalian Species 789:1–5.

Conditt, S. A., and Ribble, D. O. 1997. Social organization of *Neotoma micropus*, the southern plains woodrat. American Midland Naturalist 137:290–297.

Dial, K. P. 1988. Three sympatric species of *Neotoma*: dietary specialization and coexistence. Oecologia 76:531–537.

Escherich, P. C. 1981. The social biology of the bushy-tailed woodrat, *Neotoma cinerea*. University of California Publications in Zoology 110:1–132.

Fahrig, L., and Merriam, G. 1985. Habitat patch connectivity and population survival. Ecology 66:1762–1768.

Ford, W. M., Castleberry, S. B., Mengak, M. T., Rodrigue, J. L., Feller, D. J., and Russell, K. R. 2006. Persistence of Allegheny woodrats *Neotoma magister* across the mid-Atlantic Appalachian Highlands landscape, USA. Ecography 29:1–10.

Gerber, L. R., Seabloom, E. W., Burton, R. S., and Reichman, O. J. 2003. Translocation of an imperiled woodrat population: integrating spatial and habitat patterns. Animal Conservation 6:309–316.

Goertz, J. W. 1970. An ecological study of *Neotoma floridana* in Oklahoma. Journal of Mammalogy 51:94–104.

Hall, J. S. 1985. Eastern woodrat. *In* Species of Special Concern in Pennsylvania, eds. H. H. Genoways and F. J. Brenner, pp. 362–365. Carnegie Museum of Natural History, Pittsburgh, Pennsylvania.

Hanski, I. 1996. Metapopulation ecology. *In* Population Dynamics in Ecological Space and Time, eds. O. E. Rhodes, Jr., R. K. Chesser and M. H. Smith, pp. 13–43. University of Chicago Press, Chicago, Illinois.

HaySmith, L. 1995. *Neotoma floridana floridana*: natural history, populations, and movements in north-central Florida. Bulletin of the Florida Museum of Natural History 38:211–243.

Hickling, G. J. 1987. Seasonal reproduction and group dynamics of bushy-tailed woodrats, (*Neotoma cinerea*). Ph.D. Dissertation, University of Western Ontario, London, Ontario, Canada.

Holmes, E. E. 1995. Spatial models in ecology: explorations into the impact of social behavior on population dynamics. Ph.D. Dissertation, University of Washington, Seattle, Washington.

Hornsby, B. S., Ruiz, A. M., Castleberry, S. B., Castleberry, N. L., Ford, W. M., and Wood, P. B. 2005. Fall movements of Allegheny woodrats in harvested and intact stands in West Virginia. Northern Journal of Applied Forestry 22:281–284.

Kelly, P. A. 1989. Population ecology and social organization of dusky-footed woodrats, *Neotoma fuscipes*. Ph.D. Dissertation, University of California, Berkeley, California.

Kinsey, K. P. 1976. Social behaviour in confined populations of the Allegheny woodrat, *Neotoma floridana magister*. Animal Behaviour 24:181–187.

Kinsey, K. P. 1977. Agonistic behavior and social organization in a reproductive population of Allegheny woodrats, *Neotoma floridana magister*. Journal of Mammalogy 58:417–419.

LoGiudice, K. 2001. Latrine foraging strategies of two small mammals: implications for the transmission of *Baylisascaris procyonis*. American Midland Naturalist 146:369–378.

McGowan, E. M. 1993. Experimental release and fate study of the Allegheny woodrat (*Neotoma magister*). Federal aid project report W-166-E-1, New York Department of Environmental Conservation, Delmar, New York.

Mengak, M. T. 2002. Reproduction, juvenile growth and recapture rates of Allegheny woodrats (*Neotoma magister*) in Virginia. American Midland Naturalist 148:155–162.

Merkelz, R., and Kerr, S. F. 2002. Demographics, den use, movements, and absence of *Leishmainia mexicana* in southern plains woodrats (*Neotoma micropus*). Southwestern Naturalist 47: 70–77.

Morris, D. W. 2001. Learning from the games animals play: using behavior to assess spatial structure and stochasticity in natural populations. Annales Zoologica Fennici 38:37–53.

Moses, R. A., and Millar, J. S. 1992. Behavioural asymmetries and cohesive mother-offspring sociality in bushy-tailed wood rats. Canadian Journal of Zoology 70:57–604.

Newcombe, C. L. 1930. An ecological study of the Allegheny cliff rat (*Neotoma pennsylvanica* Stone). Journal of Mammalogy 11:204–211.

Newton, M. A. 1990. The ecology, behavior and evolutionary dynamics of the white-throated woodrat (*Neotoma albigula*). Ph.D. Dissertation, Arizona State University, Tempe, Arizona.

Poole, E. 1940. A life history sketch of the Allegheny woodrat. Journal of Mammalogy 21: 249–270.

Terborgh, J. 1974. Preservation of natural diversity: the problem of extinction prone species. BioScience 24:715–722.

Topping, M. G., and Millar, J. S. 1996a. Home range size of bushy-tailed woodrats, *Neotoma cinerea*, in southwestern Alberta. Canadian Field Naturalist 110:351–353.

Topping, M. G., and Millar, J. S. 1996b. Spatial distribution in the bushy-tailed wood rat (*Neotoma cinerea*) and its implications for the mating system. Canadian Journal of Zoology 74: 565–569.

White, G. C., and Garrott, R. A. 1990. Analysis of Wildlife Radiotracking Data. Academic Press, San Diego, California.

Wright, J. 1998. Radiotelemetry study of movements and resource use by Allegheny woodrats (*Neotoma magister*) in Pennsylvania. Project report to the Wild Resource Conservation Fund, Harrisburg, Pennsylvania.

Wright, J., and C. Butchkoski. 2005. Species account – Allegheny woodrat, *Neotoma magister*. Appendix 3. Pennsylvania's Wildlife Action Plan. Pennsylvania Game Commission, Harrisburg, Pennsylvania.

Chapter 6
Food Selection and Caching Behavior

Nikole L. Castleberry and Steven B. Castleberry

Introduction

Most species of *Neotoma* are generalist herbivores capable of adapting their feeding habits to local conditions and available food resources (Rainey 1956). Although Allegheny woodrats (*N. magister*) are considered generalist herbivores, they differ from most other woodrats as habitat specialists (Newcombe 1930, Poole 1940). Because Allegheny woodrats depend on rocky habitats that may be limiting in the surrounding environment, they cannot easily disperse in response to habitat disturbance and may be affected more by habitat change than other woodrat species (Castleberry et al. 2001). Severe habitat disturbances could cause localized extirpations, whereas similar disturbances in other woodrat species may have little effect. Because several factors affecting the availability of food resources have been suggested as contributing to localized Allegheny woodrat population declines (LoGiudice Chapter 2), knowledge of food habits and how these habits change in response to disturbance is important for conservation of the species.

Given the need for a concise source of information, we review the literature and provide a synthesis of the current knowledge of Allegheny woodrat food habits. A necessary part of the discussion is caching behavior and how caching influences food habits throughout the year. In addition, we discuss the realized and potential consequences of alterations in food resource availability from habitat disturbance.

Description of Food Habits

Several methods have been employed to determine food habits of Allegheny woodrats. Limited data have been gathered from direct observations, identification of cached items in middens assumed to be food items, and examination of stomach

N.L. Castleberry
Georgia Department of Natural Resources, Nongame Conservation Section, Social Circle, GA 30025

J.D. Peles, J. Wright (eds.), *The Allegheny Woodrat*, DOI: 10.1007/978-0-387-36051-5_6,

contents (Poole 1936, 1940, Heisler 1941, Pearson 1952, Rainey 1956). The most extensive data on food habits have been from microhistological examination of fecal material (Castleberry et al. 2002, Parker 2006).

Early observations of the Allegheny woodrat from the 1930s through the 1950s include descriptions of food habits, but are geographically limited to Pennsylvania. Fern fronds (*Polypodium vulgare*, *Pteridium latiusculum*, *Dryopteris singulosa intermedia*, and *Polystichum achrostichoides*), black birch (*Betula nigra*), eastern hemlock (*Tsuga canadensis*) and striped maple (*Acer pensylvanicum*) twigs, and birch leaves were common items found in and around nests of Allegheny woodrats and assumed to be food items (Poole 1940, Heisler 1941, Sands 1951; Table 6.1). The most common food items found by both Poole (1940) and Heisler (1941) from the viscera of freshly collected Allegheny woodrats included polypody fern (Polypodiaceae) fronds, acorns (*Quercus* spp.), American basswood (*Tilia americana*), black gum (*Nyssa sylvatica*), partidgeberry (*Michella repens*), and Juneberry (*Amelanchier canadensis*) leaves, mountain maple (*A. spicatum*), teaberry (*Gaultheria procumbens*), and white pine (*Pinus strobus*) shoots, and grape (*Vitis* spp.) fruits and seeds (Table 6.1). Other potential food items have been described from feeding observations of captive woodrats (Poole 1936, 1940, Pearson 1952, Rainey 1956).

Studies using microhistological techniques to examine seasonal food habits of Allegheny woodrats have been conducted in the Allegheny Plateau physiographic province of West Virginia, the Ridge and Valley province of West Virginia and adjacent areas in Virginia (Castleberry et al. 2002), and the Cumberland Plateau in Tennessee (Parker 2006). Castleberry et al. (2002) identified 147 total food items across the Allegheny Plateau and Ridge and Valley physiographic provinces (Table 6.1). The top year-round food items were fungi, blackberry (*Rubus* spp.) leaves, acorns, shield or wood ferns (could not be distinguished microhistologically), holly (*Ilex* spp.) or blueberry (*Vaccinium* spp.) fruit (could not be distinguished microhistologically), and lichens. In fall, woodrats relied most heavily on fungi, acorns, blackberry leaves, greenbrier (*Smilax* spp.) fruits and leaves, wood or shield fern, and oak leaves. In winter, the top five food items were ferns (*Dryopteris* spp. and *Polystichum acrostichoides*), blackberry leaves, blueberry leaves, fungi, and eastern hemlock needles. In spring, fungi were the most abundant food in the diet, followed by beech (*Fagus grandifolia*) or birch buds (could not be distinguished microhistologically), blackberry leaves, holly or blueberry fruit, and greenbrier fruit. Fungi were the most frequently used food item in summer followed by lichen, acorns, blueberry leaves, and holly or blueberry fruit. In the Cumberland Plateau, Parker (2006) documented that the summer diet was primarily forbs, shrubs, fungi, soft mast, and insects. Hard mast was the least consumed plant group comprising only 0.26% of the total diet. In winter, fungi were the main food item comprising 26.1% of the diet. Forbs, shrubs, hard mast, and insects were the primary constituents of the rest of the winter diet.

Studies on Allegheny woodrat food habits throughout the species' range have consistently identified hard and soft mast, green vegetation, fungi, and ferns as important dietary components. These results confirm the generalist diet of the species and its ability to take advantage of seasonally available foods. Several

Table 6.1 List of potential food items of Allegheny woodrats (*Neotoma magister*) from direct observations (O), found in middens (M), taken from fresh visceral samples (V), or from microhistological identification of fecal samples (F). List taken from Poole $(1940)_1$, Heisler $(1941)_2$, and Castleberry $(2000)_3$

Category/Scientific name	Common name	Leaves	Flower	Seeds	Shoots	Buds	Cones	Fruit	Bark	Twigs	Other	Unspecified
Animal												
	Unknown bones										M_2	
	White-tailed deer scat										M_2	
	Porcupine scat										M_2	
	Horse scat										M_2	
	Fox scat										M_2	
	Bird egg shells										M_2	
	Feathers										M_2	
	Woodrat hair										V_2	
	Mouse hair										V_2	
	Snail shells										M_2	
	Rabbit hair										V_2	
Fern												
Aspidium spp. (sic)	Fern	M_2										
Dryopteris marginalis	Marginal wood fern	F_3										M_1
Fern												
D. singulosa intermedia	Common wood fern											M_1
Polypodium virginianum	Polypody fern	M_2, V_2										M_1
Polystichum acrostichoides	Christmas fern	F_3										M_1
Pteridum aquillinum	Bracken fern	F_3										M_1
Fungi												
Boletus spp.	Mushroom											M_1
Boletus pallidus	Mushroom											M_1
Clavaria sp.	Coral fungus											M_1
Gautieria morchelliformis												F_3
Russula sp.												M_1
Unknown Lycoperdales (sic)	Puffballs and earthstars											V_1, F_3
Unidentified Polyporaceae	Bracket fungi											F_3
Unknown fungi	Unknown											F_3

Table 6.1 (continued)

Category/Scientific name	Common name	Leaves	Flower	Seeds	Shoots	Buds	Cones	Fruit	Bark	Twigs	Other	Unspecified
Herbaceous plant												
Acalypha sp.	Three-seeded mercury											M_1
Ambrosia artemisiifolia	Ragweed											M_1
Aralia hispida	Bristly sarsaparilla							M_1		M_1		
Aster spp.	Aster	M_1, F_3						M_1				
Carex spp.	Sedge				F_3							
Cynoglossum officinale	Gypsyflower											M_1
Epigaea repens	Trailing arbutus	V_2										
Eupatorium urticaefolium	White snakeroot											M_1
Euphorbia commutate	Tinted woodland spruge											M_1
Gaultheria procumbens	Spotted wintergreen	M_2, V_2			M_2, V_2							
Michella repens	Partridgeberry	V_2						V_2				
Oenothera biennis	Common evening primrose											M_1
Herbaceous plant												
Oxalis spp.	Wood sorrel	F_3										
Phlox brittonii	Phlox											M_1
Phytolacca americana	American beautyberry											M_1
Prenanthes spp.	Wood lettuce		F_3	F_3	M_2							
Ribes spp.	Wild currant				M_2							M_1
Solidago caesia	Wreath goldenrod										M_1	
Lichen												
Lasallia papulosa	Blistered naval lichen											M_1
Parmelia spp.	Foliose lichen											M_1
Unknown lichen	Unknown											M_2, F_3
Moss												
Dicranum sp.	Broom moss											M_1
Unknown moss	Unknown											F_3
Unknown plant												
Unknown flowers	Flowers		F_3									
Unknown leaves	Leaves	F_3										
Unknown seeds	Seeds			F_3								

Table 6.1 (continued)

Category/Scientific name	Common name	Leaves	Flower	Seeds	Shoots	Buds	Cones	Fruit	Bark	Twigs	Other	Unspecified
Shrub												
Amelanchier canadensis	Serviceberry	V_2										M_1
Aralia spinosa	Devil's walking stick							F_3				
Crataegus spp.	Hawthorn	F_3										
Elaeagnus umbellate	Autumn olive	F_3		F_3								
Illex spp.	Holly			F_3				F_3				
Kalmia latifolia	Mountain laurel				M_2							
Rhododendron maximum	Rosebay rhododendron	V_1, F_3								M_1		
Rhus typhina	Staghorn sumac											M_1
Sambucus canadensis	Elderberry	M_1								M_1		
Sambucus spp.	Elderberry	F_3										
Vaccinium spp.	Blueberry	F_3		F_3				F_3				
Viburnum spp.	Virburnum							F_3				
Tree												
Acer pensylvanicum	Striped maple				M_2							
Acer spicatum	Mountain maple	M_2			M_2, V_2							
Alnus serrulata	Smooth alder	F_3										
Betula spp.	Unknown birch	F_3			M_2	F_3						
Tree												
Betula lenta	Black birch			V_1						V_1		M_1
Carya alba	Mockernut hickory			M_1								
Carya glabra	Red hickory											M_1
Carya spp.	Unknown hickory	F_3		F_3								
Castanea dentata	American chestnut	M_2		V_1	M_2						M_1	
Cercis canadensis	Redbud	F_3		F_3								M_1
Cornus florida	Flowering dogwood							V_1				
Cornus spp.	Dogwood	M_2, F_3			M_2							
Fagus grandifolia	American beech	F_3				F_3						
Fraxinus americana	White ash											M_1
Juglans nigra	Black walnut			F_3								M_1
Juniperus virginiana	Eastern red cedar								M_1			
Magnolia fraseri	Mountain magnolia	F_3		F_3			F_3					

Table 6.1 (continued)

Category/Scientific name	Common name	Leaves	Flower	Seeds	Shoots	Buds	Cones	Fruit	Bark	Twigs	Other	Unspecified
Malus pumila	Apple			V_1				M_1, V_1				
Nyssa sylvatica	Blackgum	V_2										
Pinus strobes	White pine				V_2							
Prunus spp.	Cherry	F_3										
Tree												
Prunus pennsylvanica	Fire cherry	M_1										M_1
Prunus serotina	Black cherry			V_1				V_1				M_1
Quercus coccinia	Southern red oak									M_1		
Quercus ilicifolia	Bear oak			V_1								
Quercus montana	Chestnut oak	M_2		V_2								
Quercus muhlenbergii	Chinkapin oak	M_1	M_1						M_1			
Quercus rubra	Northern red oak	M_1		M_1								
Quercus spp.	Unknown oak	F_3		F_3								
Quercus velutina	Black oak											M_1
Robinia pseudoacacia	Black locust											
Sassafras albidum	Sassafras	M_1								M_1		
Sorbus americana	Mountain ash			V_1				V_1				
Tilia glabra	Basswood	V_2										
Tsuga canadensis	Hemlock	M_2, F_3		V_1	M_2					M_1, V_1		
Vine												
Parthenocissus quinquefolia	Virginia creeper				M_2							
Rubus spp.	Blackberry	M_2, F_3			M_2, V_2			F_3				
Smilax spp.	Greenbrier	F_3		F_3				F_3				
Unidentified Roseaceae	Rose	F_3		F_3								
Vitis aestivalis	Summer grape											M_1
Vitis spp.	Wild grape			V_2				V_2				

studies suggest that generalists can assess the nutrient value of foods and select a diet that most readily meets nutritional needs (Lindloff et al. 1974, Glander 1981, Eschelman and Jenkins 1989, Post 1993).

Acorn availability is an important factor affecting the growth and survival of young and maintaining body weights of adults in many rodents, such as white-footed mice (*Peromyscus leucopus*; Batzli 1977, Hansen and Batzli 1978) and gray squirrels (*Sciurus carolinensis*; Nixon et al. 1968, 1975). Although Allegheny woodrats consume the greatest quantities of hard mast in the late summer and throughout the fall, hard mast is consumed in every season (Castleberry et al. 2002, Parker 2006). A similar pattern of hard mast consumption has been observed in the eastern woodrat (*N. floridana*; Wagle and Feldhamer 1997). Additionally, acorn abundance in the diet of the Allegheny woodrat tends to track annual production and availability. Hard mast consumption in summer is likely due to early acorn drops, although woodrats are capable of climbing trees (Risch and Brady 1996) and may eat unripe acorns as early as July. Significant acorn use in areas such as the Allegheny Plateau, where oak-dominated forest stands are rare, highlights the importance of hard mast to Allegheny woodrats (Castleberry et al. 2002). Furthermore, there is evidence to suggest that Allegheny woodrat populations in the parts of the range where oak-dominated forests are predominant (e.g., Ridge and Valley province) are sensitive to episodic production of acorns (Castleberry et al. 2002).

Many soft mast producing species represent high quality food items. Fleshy fruits are similar to acorns in nutrient and digestibility values (Short and Epps 1976). Generally, soft mast use is highest during spring and summer (Castleberry et al. 2002, Parker 2006). However, use of soft mast and insect material may be grossly underestimated in diets studied using microhistology due to high rates of digestion (Litvaitis et al. 1994).

Fungi represent another high quality food category that has been consistently documented in Allegheny woodrat diets. Castleberry et al. (2002) found that fungi represented ≥9.7% of the total diet in every season in the Allegheny Plateau. Similarly, Parker (2006) found fungi to represent 12.7% and 21.6% of the summer and winter diets, respectively, in the Cumberland Plateau. Forest communities dominated by trees with ectomycorrhizal associations like Fagaceae or Betulaceae, such as in the Allegheny and Cumberland Plateaus, produce a greater abundance of relatively large, fleshy mushrooms and hypogeous fungi than many other forest types (Trappe 1962). Additionally, the mesic conditions found in the Allegheny Plateau facilitate mushroom growth. Compared to pine mast and nuts such as beechnuts, chestnuts (*Castanea dentata*), walnuts (*Juglans nigra*), and hickory nuts (*Carya* spp.), dried mushrooms are richer in protein, carbohydrates, and minerals (Short and Epps 1976, Fogel and Trappe 1978). Additionally, dried fungi preserve well, making them well-suited for caching. The effect of poor mast crops, traditionally considered a key food item in many small mammal diets in the Appalachians, may be moderated in years with a large mushroom crop (Baumgartner 1939, Brown and Yeager 1945, Lampio 1967, Ognev 1940, Rajala and Lampio 1963, Stienecker and Browning 1970, Uhlig 1955, Fogel and Trappe 1978). Nonetheless, caution must be used when making inferences about fungi presence in Allegheny woodrat diets.

Eastern woodrats have been shown to choose foods highly infected with fungal growth (Herrera and McDonald 1997). In fact, they tend to select highly fungus-infected foods over ones with intermediate fungal infections. These foods may have increased nutritional value from fungal breakdown of unavailable nutrients into usable compounds or are more attractive due to fungal degradation of toxic secondary compounds (Rossi 1985, Petersen et al. 1989).

Green vegetation is consumed in high amounts ($\geq$15% of the total diet) in all seasons throughout much of the range (Castleberry 2000, Parker 2006). Although Castleberry et al. (2002) found differences in consumption of green vegetation among seasons for both the Ridge and Valley and Allegheny Plateau physiographic provinces, total food item and species richness in the diet generally was highest during the winter and spring when woodrats rely primarily on cached foods. Both the amount and richness of green plants consumed were generally lower during summer and fall when food availability was highest. Green vegetation also appears to be an important buffer food in the Ridge and Valley and Cumberland Mountains when mast availability is low (Castleberry et al. 2002, Parker 2006).

The importance of arthropods in the diet, particularly soft-bodied coleopteran and lepidopteran larvae, has been documented for several small mammal species (Hamilton 1941, Whitaker 1966, Nixon et al. 1968, Alcoze and Zimmerman 1973, Houtcooper 1978, Maser et al. 1985). Several species of woodrats have been reported to feed on animal matter, chiefly insects, though it has never been reported as a dominant food item (Poole 1940, Vorhies and Taylor 1940, Linsdale and Tevis 1951, Murphy 1952, Fitch and Rainey 1956, Horne 1996, Williams et al. 2000). Allegheny woodrats have been shown to consume insect material year-round (Castleberry et al. 2002, Parker 2006). In the Ridge and Valley of West Virginia, insects made up >5% of the summer diet and >3% of the fall diet, whereas they only comprised approximately 1% of the summer diet and 3% of the fall diet in the Allegheny Plateau (Castleberry 2000). Female woodrats used almost twice as much insect material in the summer and males used almost twice as much in the fall (Castleberry 2000). These differences are likely due to differences in nutritional needs, primarily protein, between the sexes. However, as noted above, reports of insect consumption likely are underestimates of the actual amount consumed (Litvaitis et al. 1994).

Caching Behavior

Allegheny woodrats forage in the vicinity of the nest in spring and summer (Castleberry et al. 2001), but rely on cached food during portions of the year when other foods are rare or unavailable. Caching begins in September or October and dominates all other activities until the onset of winter (N. Castleberry pers. obs.). Cached materials are the primary food source during winter but limited foraging still occurs, presumably to avoid being left with poor quality foods before the cache can be restocked. Woodrats often store more food than they can consume in winter such that leftover material is still present when caching begins in the next growing season.

Similar to other species that exhibit caching behavior, Allegheny woodrats are faced with both long- and short-term foraging decisions (Katz 1974, Pyke et al. 1977, Craig et al. 1979). According to optimal foraging theory (Reichman and Fay 1983) woodrats would be expected to have lower total food item richness and diversity values in seasons of high food resource abundance than when relying on cached foods. When foraging for the short term (i.e., when food resources are abundant), animal species that cache should consume only food items that are highly preferred and would fulfill daily nutritional and energy requirements. This behavior prevents those resources from being acquired by other individuals (Reichman and Fay 1983). When relying on cached foods, animals would be expected to consume a more diverse diet to ensure they have access to high quality food items at the end of the cache-dependent period (Reichman and Fay 1983). Observed food habits of the Allegheny woodrat appear to follow the predicted patterns of optimal foraging strategies. Castleberry et al. (2002) found that species richness values in the diet were significantly higher in winter when woodrats were relying on cached food items.

In addition to food items, Allegheny woodrats also commonly cache nonfood items, such as human refuse, feathers, bones, and scats of other animals (Poole 1940). Although the purpose of caching nonfood items is unknown, woodrats consume seeds contained in the fecal material of other species, especially raccoons (*Procyon lotor*; LoGiudice 2001). Caching raccoon feces exposes woodrats to the eggs of the raccoon roundworm (*Baylisascaris procyonis*), a highly pathogenic nematode that has been implicated as a contributing factor in the decline of the species in the northern parts of the distribution (LoGiudice 2001, 2003, 2006). Allegheny woodrats are more susceptible to raccoon roundworm infection than other small mammals because they only cache dried scat which allows the parasite eggs to embryonate into the pathogenic stage (LoGiudice 2001). Furthermore, they store feces on food caches potentially contaminating the cache with parasite eggs.

Disturbance of Food Resources as Potential Factors for Population Decline

Of the numerous factors suggested as contributing to the decline of Allegheny woodrat populations, several relate to changes in food resource availability resulting from anthropogenic influences. For example, given the importance of hard mast, reduction of annual hard mast crops resulting from chestnut blight (*Cryphonectria parasitica*) or gypsy moth (*Lymantria dispar*) infestations of oak stands (Hall 1988) may have adverse affects on populations. Neal (1967) found a direct correlation between mast failures and a rapid decline in densities of *N. floridana* in Louisiana.

Wright and Kirkland (2000) present circumstantial evidence linking woodrats declines to the disappearance of the American chestnut. They illustrate that the distributional ranges of the Allegheny woodrat and American chestnut overlap and that Allegheny woodrat declines followed chestnut suppression both geographically and temporally. Wright and Kirkland (2000) also speculated that chestnuts may have been essential in Allegheny woodrat nutrition and that acorns are an inadequate substitute. Although acorns are higher in lipids and calcium (United States Department of Agriculture 1999), annual acorn production is highly variable, making them a less reliable food source than the American chestnut that exhibited less annual variation. LoGiudice (2006) speculated that annual variation in acorn crops may lead to variable reproductive output in Allegheny woodrats. Furthermore, the higher lipid content makes acorns more perishable than chestnuts while stored in winter caches.

Changes in plant species composition due to a variety of factors may affect food resources and contribute to Allegheny woodrat declines (Balcom and Yahner 1996). Although the impact of white-tailed deer (*Odocoileus virginianus*) herbivory on Allegheny woodrat food availability is unknown, overabundant deer populations have profound influences on plant species composition and structure (Alverson 1988, Tilghman 1989, Horsley et al. 2003). In particular, deer herbivory can prevent regeneration of oaks and other mast producing species (Marquis 1974, Marquis and Grisez 1978). Similarly, historic fire suppression throughout the eastern hardwood region has created conditions favorable to faster-growing species such as birch and maples at the expense of slower-growing, more fire-tolerant oaks (Spurr and Barnes 1980, Abrams and Nowacki 1992). The alteration of food availability by the combination of high deer densities and fire suppression throughout much of the range may have played a key role in woodrat population declines.

Conservation and Management of Declining Species

Knowledge of food habits is vital for successful long-term management of most wildlife species (Bump et al. 1947, Korschegen 1962, 1967, Goodrum et al. 1971, Nixon et al. 1975, Pelton 1983, Nixon and Hansen 1987). In this chapter, we have provided an overview of Allegheny woodrat food habits and how populations may be affected by changes in food resource availability. The extent to which knowledge of Allegheny woodrat food habits and responses to changes in food availability applies to other species depends on numerous species-specific factors. Nonetheless, there are some points for consideration that generally apply to declining species.

The foraging behavior of the Allegheny woodrat may make inferences particularly pertinent to other food hoarding species. Management for food hoarding species may differ from other species because they rely primarily on cached foods for part of the year. Therefore, a primary consideration in managing for food hoarding species is identifying and promoting foods that have the appropriate characteristics for long-term storage. Eastern woodrats are capable of discriminating between food items based on perishability (Reichman 1988). For example, Allegheny woodrat food habits studies have shown that hard mast is stored for future consumption and found in the diets year-round whereas soft mast is more perishable and is only found in the diet in summer and fall when fresh fruits are available. Conversely, a primary consideration for management for noncaching species is year-round food availability. Species that do not cache food must find fresh food year-round or store enough energy to get through times of limited resource availability. Furthermore, because many noncaching species do not have the benefit of feeding from the cache in relative safety, foraging cover may be a more important habitat component.

Providing a diversity of food items may also be an important management consideration for declining species. Castleberry et al. (2002) suggested promoting traditional mast species such as oak, cherry, and beech, as well as nontraditional mast species such as grape, blueberry, maple, and greenbrier to minimize the effects of mast availability variation on Allegheny woodrats. Promoting these nontraditional species as buffer species provides mast in seasons when traditional mast is not available (Edwards et al. 1993). Furthermore, buffer species compensate during years of poor traditional mast production.

Reduction in food availability as a result of the effects of the gypsy moth and chestnut blight likely has played a role in Allegheny woodrat declines in parts of the range (Balcom and Yahner 1996, LoGiudice 2006). Although this hypothesis is largely supported by circumstantial evidence (Wright and Kirkland 2000), the impacts of invasive, exotic pests on declining species require attention. Numerous examples of deleterious effects of exotic species on native species have been documented. As with the Allegheny woodrat, exotic species may seem inconsequential as single factors, but may interact with other factors to produce serious negative consequences. Therefore, eradication or control of invasive species may be critical for long-term conservation of many declining species.

References

Abrams, M. D., and Nowacki, G. J. 1992. Historical variation in fire, oak recruitment, and post-logging accelerated succession in central Pennsylvania. Bulletin of the Torrey Botanical Club 119: 19–28.

Alcoze, T. M., and Zimmerman, E. G. 1973. Food habits and dietary overlap of two heteromyid rodents from the mesquite plains of Texas. Journal of Mammalogy 54: 900–908.

Alverson, W. S. 1988. Forests too deer: edge effects in northern Wisconsin. Conservation Biology 2: 348–358.

Balcom, B. J., and Yahner, R. H. 1996. Microhabitat and landscape characteristics associated with the threatened Allegheny woodrat. Conservation Biology 10: 515–523.

Batzli, G. O. 1977. Population dynamics of the white-footed mouse in floodplain and upland forests. American Midland Naturalist 97: 18–32.

Baumgartner, L. L. 1939. Foods of the fox squirrel in Ohio. Ohio Wildlife Research Station Release 108, Ohio State University, Columbus, Ohio.

Brown, L. G., and Yeager, L. E. 1945. Fox squirrels and grey squirrels in Illinois. Illinois Natural History Survey Bulletin 23: 449–536.

Bump, G., Darrow, R. W., Edminster, F. C., and Crissey, W. F. 1947. The Ruffed Grouse: Life History, Propagation, and Management. New York State Conservation Department, Albany, New York.

Castleberry, N. L. 2000. Food habits of the Allegheny woodrat (*Neotoma magister*). M.S. Thesis, West Virginia University, Morgantown, West Virginia.

Castleberry, S. B., Ford, W. M., Wood, P. B., Castleberry, N. L., and Mengak, M. T. 2001. Movements of Allegheny woodrats in relation to timber harvesting. Journal of Wildlife Management 65: 148–156.

Castleberry, N. L., Castleberry, S. B., Ford, W. M., Wood, P. B., and Mengak, M. T. 2002. Allegheny woodrat (*Neotoma magister*) food habits in the Central Appalachians. American Midland Naturalist 147: 80–92.
Craig, R. B., DeAngelis, D. L., and Dixon, K. R. 1979. Long- and short-term dynamic optimization models with application to the feeding strategy of the loggerhead shrike. American Naturalist 113:31–51.
Edwards, J. W., Guynn, D. C., Jr., and Loeb, S. C. 1993. Seasonal mast availability for wildlife in the Piedmont region of Georgia. USDA Forest Service Research Paper SE-287, Asheville, North Carolina.
Eschelman, B. D., and Jenkins, S. H. 1989. Food selection by Belding's ground squirrels in relation to plant nutritional features. Journal of Mammalogy 70: 846–852.
Fitch, H. S., and Rainey, D. S. 1956. Ecological observations of the woodrat (*Neotoma floridana*). University of Kansas Publication of the Museum of Natural History 8: 499–533.
Fogel, R. D., and Trappe, J. M. 1978. Fungus consumption (mycophagy) by small mammals. Northwest Scientist 52: 1–31.
Glander, K. E. 1981. Feeding patterns in mantled howler monkeys. *In* Foraging Behavior: Ecological, Ethological, and Psychological Approaches, eds. A. C. Kamil and T. C. Sargent, pp. 231–257. Garland STPM Press, New York.
Goodrum, P. D., Reid V. H., and Boyd, C. E. 1971. Acorn yields, characteristics, and management criteria of oaks for wildlife. Journal of Wildlife Management 35: 520–532.
Hall, J. S. 1988. Survey of the woodrat in Pennsylvania. Contracts 878903 and 878919, Final report to the Pennsylvania Game Commission, Harrisburg, Pennsylvania.
Hamilton, W. J., Jr. 1941. The food of small forest mammals in the Eastern United States. Journal of Mammalogy 21: 250–263.
Hansen, L., and Batzli, G. O. 1978. The influence of food availability on the white-footed mouse: populations in isolated woodlots. Canadian Journal of Zoology 56: 2530–2541.
Heisler, W. T. 1941. Allegheny woodrat populations. M.S. Thesis, Pennsylvania State College, University Park, Pennsylvania.
Herrera, J., and McDonald, M. W. 1997. Consumption by eastern woodrats (*Neotoma floridana*) of food infected by fungi. American Midland Naturalist 137: 282–289.
Horsley, S. B., Stout, S. L., and DeCalesta, D. S. 2003. White-tailed deer impact on the vegetation dynamics of a northern hardwood forest. Ecological Applications 13: 98–118.
Horne, E. A. 1996. Construction, use, and defense of a central food cache by the eastern woodrat, *Neotoma floridana*. Ph.D. Dissertation, Kansas State University, Manhattan, Kansas.
Houtcooper, W. C. 1978. Food habits of rodents in a cultivated ecosystem. Journal of Mammalogy 59: 427–430.
Katz, P. L. 1974. A long-term approach to foraging optimization. American Naturalist 108: 758–782.
Korschegen, L. J. 1962. Foods of Missouri deer, with some management implications. Journal of Wildlife Mangement 26: 164–172.
Korschegen, L. J. 1967. Feeding habits and foods. *In* The Wild Turkey and its Management, ed. O. H. Hewitt, pp. 137–138. Wildlife Society, Washington, D.C.
Lampio, T. 1967. Sex ratios and the factors contributing to them in the squirrel, *Sciurus vulgaris*, in Finland II. Finnish Game Research 29: 5–69.
Lindloff, B., Lindstrom, E., and Pehrson, A. 1974. Nutrient content in relation to food preferred by mountain hare. Journal of Wildlife Management 38:875–879.
Linsdale, J. M., and Tevis, L P., Jr. 1951. The Dusky-Footed Woodrat. University of California Press, Berkeley, California.
Litvaitis, J. A., Titus, K., and Anderson, E. M. 1994. Measuring vertebrate use of terrestrial habitats and foods. *In* Research and Management Techniques for Wildlife and Habitats, Fifth edition, ed. T. A. Bookhout, pp. 254–306. The Wildlife Society, Bethesda, Maryland.
LoGiudice, K. 2001. Latrine foraging strategies of two small mammals: implications for the transmission of *Baylisascaris procyonis*. American Midland Naturalist 146: 369–378.

LoGiudice, K. 2003. Trophically transmitted parasites and the conservation of small populations: raccoon roundworms and the imperiled Allegheny woodrat. Conservation Biology 17: 258–266.

LoGiudice, K. 2006. Toward a synthetic view of extinction: a history lesson from a North American rodent. Bioscience 56: 687–693.

Marquis, D. A. 1974. The impact of deer browsing on Allegheny hardwood regeneration. USDA Forest Service Research Paper NE-308, Upper Darby, Pennsylvania.

Marquis, D. A., and Grisez, T. J. 1978. The effect of deer exclosures on the recovery of vegetation in failed clearcuts on the Allegheny plateau. USDA Forest Service Research Paper NE-270, Upper Darby, Pennsylvania.

Maser, Z., Maser, C., and Trappe, J. M. 1985. Food habits of the northern flying squirrel (*Glaucomys sabrinus*) in Oregon. Canadian Journal of Zoology 63: 1084–1088.

Murphy, M. F. 1952. Ecology and helminths of the osage wood rat, *Neotoma floridana osagensis*, including the description of the *Longistriata neotoma* n. sp. (Trichostoronblidae). American Midland Naturalist 48: 204–218.

Neal, W. A. 1967. A study of the ecology of the woodrat in the hardwood forests of the lower Mississippi river basin. M.S. Thesis, Louisiana State University, Baton Rouge, Louisiana.

Newcombe, C. L. 1930. An ecological study of the Allegheny wood rat. Journal of Mammalogy 11: 204–211.

Nixon, C. M., and Hansen, L. P. 1987. Managing forests to maintain populations of gray and fox squirrels. Illinois Department of Conservation Technical Bulletin 5, Springfield, Illinois.

Nixon, C. M., Worley, D. M., and McClain, M. W. 1968. Food habits of squirrels in southeast Ohio. Journal of Wildlife Management 32: 294–305.

Nixon, C. M., McClain, M. W., and Donohoe, R.W. 1975. Effects of hunting and mast crops on a squirrel population. Journal of Wildlife Management 39: 1–25.

Ognev, S. I. 1940. Mammals of the USSR and Adjacent Countries. Volumes IV–VII. Translated By Jean Salkind. Smithsonian Institute and National Science Foundation, Washington, D.C.

Parker, W. T. 2006. Immobilization of small mammals and occupancy, seasonal food habits, and parasites of Allegheny woodrats in the Cumberland mountains, Tennessee. M.S. Thesis, University of Tennessee, Knoxville, Tennessee.

Pearson, P. G. 1952. Observations concerning the life history and ecology of the woodrat, *Neotoma floridana floridana* (Ord). Journal of Mammalogy 33: 459–463.

Pelton, M. R. 1983. Movements. *In* Tri-State Black Bear Study, eds. D. M. Carlock, R. H. Conley, J. M. Collins, P. E. Hale, K. G. Johnson, A. S. Johnson and M. R. Pelton, pp. 134–180. Tennessee Wildlife Resources Agency Technical Report Number 83-9, Nashville, Tennessee.

Petersen, R. C., Cummins, K. W., and Ward, G. M. 1989. Microbial and animal processing of detritus in a woodland stream. Ecological Monographs 59: 21–39.

Poole, E. L. 1936. Notes on the young of the Allegheny wood rat. Journal of Mammalogy 17: 22–26.

Poole, E. L. 1940. A life history sketch of the Allegheny woodrat. Journal of Mammalogy 21: 249–270.

Post, D. M. 1993. Detection of differences in nutrient concentrations by eastern woodrats (*Neotoma floridana*). Journal of Mammalogy 74: 493–497.

Pyke, G., Pulliam, H. R., and Charnov, E. L. 1977. Optimal foraging: a selective review of theory and tests. Quarterly Review of Biology 52: 137–154.

Rainey, D. G. 1956. Eastern woodrat, *Neotoma floridana*: life history and ecology. University of Kansas Museum of Natural History Publication 8: 369–375.

Rajala, P., and Lampio, T. 1963. The food of the squirrel (*Sciurus vulgaris*) in Finland in 1945–1961. Suomen Riisata 16: 155–185.

Reichman, O. J. 1988. Caching behavior of eastern woodrats, *Neotoma floridana*, in relation to food perishability. Animal Behaviour 36: 1525–1532.

Reichman, O. J., and Fay, P. 1983. Comparison of the diets of a caching and non-caching rodent. American Naturalist 122: 576–581.

Risch, T. S., and Brady, M. J. 1996. Trap height and capture success of arboreal small mammals: evidence from southern flying squirrels (*Glaucomys volans*). American Midland Naturalist 136: 346–351.
Rossi, L. 1985. Interactions between invertebrates and microfungi in freshwater ecosystems. Oikos 44: 175–184.
Sands, D. E. 1951. A Study of the Allegheny woodrat in central Pennsylvania. M.S. Thesis, Pennsylvania State College, University Park, Pennsylvania.
Short, H. L., and Epps, E. A., Jr. 1976. Nutrient quality and digestibility of seeds and fruits from southern forests. Journal of Wildlife Management 40: 283–289.
Spurr, S. H., and Barnes, B. V. 1980. Forest Ecology. Third edition. John Wiley and Sons, New York.
Stienecker, W., and Browning, B. M. 1970. Food habits of the western gray squirrel. California Fish and Game Publication 56: 36–48.
Tilghman, N. G. 1989. Impacts of white-tailed deer on forest regeneration in northwestern Pennsylvania. Journal of Wildlife Management 53: 524–532.
Trappe, J. M. 1962. Fungus associates of ectotropic mycorrhizae. Botanical Review 28: 538–606.
Uhlig, H. C. 1955. The Gray Squirrel: Its Life History, Ecology, and Population Characteristics in West Virginia. West Virginia Conservation Commission, Charleston, West Virginia.
United States Department of Agriculture. 1999. USDA Nutrient Database for Standard Reference, Release 12. Available http://www.rahul.net/cgi-bin/fatfree/usda/usda.cgi
Vorhies, C. T., and Taylor, W. P. 1940. Life history and ecology of the white-throated wood rat, *Neotoma albigula albigula* Hartley, in relation to grazing in Arizona. University of Arizona College of Agriculture Technical Bulletin 86: 455–529.
Wagle, E. R., and Feldhamer, G. A. 1997. Feeding habits of the eastern woodrat (*Neotoma floridana*) in southern Illinois. Transactions of the Illinois State Academy of Science 90: 171–177.
Whitaker, J. O., Jr. 1966. Food of *Mus musculus*, *Peromyscus maniculatus bairdi* and *Peromyscus leucopus* in Vigo County, Indiana. Journal of Mammalogy 47: 473–486.
Williams, D.K., Van Why, K., and Applegate, R. D. 2000. Eastern woodrat (*Neotoma floridana*) consumption of northern bobwhite (*Colinus virginianus*). American Midland Naturalist 142: 239–244.
Wright, J., and Kirkland, G. L., Jr. 2000. A possible role for chestnut blight in the decline of the Allegheny woodrat. Journal of the American Chestnut Foundation 8: 30–35.

Section III
Conservation and Management

The chapters in Sections I and II summarize our current knowledge of causes of decline and of basic ecology for the Allegheny woodrat (*Neotoma magister*). Using the information in those sections as a framework, the chapters in this section outline procedures and policies to implement conservation efforts for the Allegheny woodrat.

Monitoring is necessary both for the assessment of a species' status as well as for determining the success of conservation efforts. The authors of Chapter 7 represent each of the four states that have had long-term monitoring programs for the Allegheny woodrat. Mike Mengak (Virginia), Cal Butchkoski (Pennsylvania), Dan Feller (Maryland), and Scott Johnson (Indiana) highlight the accomplishments of the programs in each of their states. The authors discuss these accomplishments in the context of developing a standardized program that will represent a model for all states to use in monitoring the status of *N. magister*.

An outcome of the monitoring program in the state of Pennsylvania has been an extensive data set representing the past and present distribution of the Allegheny woodrat as well as factors related to the occurrence of woodrats. These data provide a basis for a surface-rock community management plan proposed by Jerry Hassinger, Cal Butchkoski, and Duane Diefenbach in Chapter 8. The authors include numerous recommendations for implementing this plan and stress the importance of coordinating efforts with forestry and other resource managers. Although this plan was developed for the state of Pennsylvania, it provides a model for other states, and in fact, the authors urge coordination of management efforts among states.

Although the genetics of a species, like its ecology, is part of its basic biology, this topic is addressed in this conservation section because of its important relationship to reintroductions. In Chapter 9, Tim Smyser and Gene Rhodes review our current knowledge of the genetics of *N. magister*. Studies to date reveal that gene flow is limited in this species and that several isolated populations are in danger of extirpation due to reduced genetic variability. The authors present a case study of woodrat populations in Indiana and discuss the implications of reduced genetic variability in these populations for reintroduction efforts in the state.

In addition to managing Allegheny woodrats in areas where the species is currently extant, reintroduction to regions where the species has been extirpated is another major component of conservation. As discussed by Tom Serfass in Chapter 10, reintroduction is a complex process, and several previous attempts to reintroduce Allegheny woodrats have failed. The author uses his expertise on this topic to provide a general overview of the procedures, and then makes specific recommendations for the reintroduction of woodrats to extirpated sites and supplementation of woodrats in areas where populations have declined.

In this book, we have made the case that much of what we are learning from the Allegheny woodrat is applicable to other species. Nowhere is this more true than within the genus *Neotoma* where several species or subspecies have also experienced declines in abundance and distribution. In Chapter 11, George Feldhamer and Aaron Poole review the current status of imperiled *Neotoma* in North America. The decline of several, mostly insular, species can be directly linked to specific factors such as species introductions and habitat loss. However, one subspecies of *N. floridana* (*N. f. illinoensis*) appears to be experiencing declines that are similar in complexity to the Allegheny woodrat. The authors focus specifically on the case of *N. f. illinoensis* and examine the results of a reintroduction project for that subspecies.

John D. Peles
Janet Wright

Chapter 7
Lessons from Long-Term Monitoring of Woodrat Populations

Michael T. Mengak, Calvin M. Butchkoski, Daniel J. Feller, and Scott A. Johnson

Introduction

The historic range of the Allegheny woodrat *Neotoma magister* extended from Connecticut west to Indiana and south along the Appalachian Mountains to northeast Alabama (Wright Chapter 1). The Allegheny woodrat was initially listed as a Category II species under the Endangered Species Act of 1973 by the US Fish and Wildlife Service (Federal Register 1989). While Category II designations are no longer used by regulatory agencies, the Allegheny woodrat remains imperiled throughout its geographic range (Wright Chapter 1).

Allegheny woodrat habitat consists of caves, cliffs, talus fields, rock outcrops, or other rocky substrate having crevices sufficient in number and depth to accommodate woodrats (Ford et al. 2006). By the very nature of this habitat, potential woodrat sites are widely distributed across the landscape. The intervening habitat matrix is a fragmented patchwork of forest and human modified elements. Woodrats exist across this matrix as isolated clusters. This metapopulation arrangement places woodrats and other similarly distributed species at an increased risk for extinction of regional populations (Hanski and Gilpin 1991). Thus, an effective monitoring program aimed at rapidly detecting declines in population size or distribution of population clusters is crucial if management activities are to be undertaken in time to prevent total loss of clusters or the species itself.

Status checks of woodrat locations can be divided into monitoring programs and inventory or survey studies. The goal of a monitoring program is, or should be, to detect changes in the magnitude and direction (increase or decrease) in average numbers of animals at several sites over a well-defined time period (Thompson et al. 1998). Inventory or survey studies, by contrast, provide spot-checks for the activity of woodrats at sites across the species range, once or at long temporal intervals. Each will yield different information, and the two may work best in combination.

M.T. Mengak
Warnell School of Forestry and Natural Resources, University of Georgia, Athens, GA 30602

J.D. Peles, J. Wright (eds.), *The Allegheny Woodrat*, DOI: 10.1007/978-0-387-36051-5_7,

Sampling of rare species, such as the Allegheny woodrat, entails special problems (Thompson 2004). While presence is easily confirmed with the capture of at least one individual, absence can only be deduced after repeated attempts to locate a woodrat result in failure. Time and budget constraints determine the number of repeated attempts a program can make in a single area of potential habitat before categorizing that area as unoccupied.

Accessibility of cliffs, caves, and talus/rock debris is another challenge. Private landowners may not support endangered or rare species surveys. Public lands may lack roads or trails that also make access difficult. Selection of sites for long-term monitoring (LTM) may be based on access (determined by legal, time, and monetary constraints) and not due to considerations of statistical randomness. Questions of extensive versus intensive sampling will vex the investigators seeking to balance experimental design with real-world constraints of time and money that can be devoted to sampling.

A metapopulation may be defined as a collection of local populations linked by dispersal (Hanski and Gilpin 1991). Although the Allegheny woodrat exhibits a metapopulation structure (Wood Chapter 3), the question of how to define a local population is left to individual investigators. Some population site limits are easy to define, such as a cave and its entrance, but others are more arbitrarily delineated boulder fields and scattered rock piles. *A priori* knowledge of woodrat movements from one rock field to another is often lacking, and thus what seems like distinct rock fields may actually be a single continuous area of habitat. Allowing the rock outcrop or cliff to define the extent of the population may be practical, but could give a different outcome compared to letting the animals define the population boundaries. However, the latter can only accurately occur with radiotelemetry (expensive) or genetic sampling (*a posteriori*).

Sampling protocols for point (cave), linear (cliff lines), and area (boulder fields) habitat may differ based on study objectives. Occurrence will be determined by simple presence or absence, but more complicated metrics, such as abundance or density, will require site-specific sampling protocols. There is no substitute for an actual capture to determine occurrence, but given constraints of time and money, indirect measures of occurrence may prove useful in locating potential sites for more intensive study. Indirect measures include visual surveys for feces, middens, or nests. These measures should only be used to evaluate trends in populations if validated by other independent means. Resampling should be frequent enough to detect the trends in population status and recolonization of formerly occupied habitat. Additional data collected by monitoring or surveys should include, at a minimum, information on weight, sex, relative age, and overall health of the animals. Ancillary information could include ectoparasites and tissue for genetic analysis, feces for food habitats studies, and site observations of food caches and latrines. Surrounding landscape characteristics can be gathered from ground surveys or remotely sensed data sources. A global positioning system and geographic information system (GIS) should be used to record and interpret data.

Beyond the basic ground rules, monitoring practices for a species of concern can take various forms. Here, we review ongoing monitoring and survey programs of Allegheny woodrats from Pennsylvania, Indiana, Maryland, and Virginia. These states have the longest history of monitoring woodrat populations and serve as models for monitoring programs. Insight gained through these monitoring programs has greatly advanced our knowledge of woodrat biology and ecology.

Pennsylvania Program

Acting on anecdotal cases that suggested the Allegheny woodrat was disappearing from some of its historic range in Pennsylvania, a network of mammalogists initiated a live-trapping program at several sites across the state in 1982. Trapping was done in mid-September through mid-October under favorable weather conditions with 40 live traps on two consecutive nights. Live-trapping protocols for Pennsylvania, as well as for the other states' programs described in this chapter, used 16-ga wire mesh Tomahawk live-traps (Models 102 and 201) baited with sliced apples or with rolled oats and peanut butter. Captures were ear-tagged, weighed, and reproductive condition noted. Although the raw data from the first year of this effort unfortunately did not survive, the standardized protocol that was developed, the increased knowledge of woodrat habitat, and the summary results of trapping convinced the Pennsylvania Game Commission that a more coordinated program was needed.

Beginning in 1987, four long-term monitoring sites were trapped yearly through the mid-1990s (Fig. 7.1). At one site, Wolford Gap, defoliation by gypsy moth (*Lymantria dispar*) was observed in the first year, with severe tree mortality noted in 1990. This site was supplemented with acorns from both red and white oaks (*Quercus* spp.) and a 50/50 mixture of oats and wheat, three times each year in fall, winter, and spring from 1988 to 1993. By 1990, the Pennsylvania Game Commission established a primary objective to complete an inventory of active woodrat sites statewide. To accomplish this task, survey priorities changed from labor-intensive live-trapping to visual identification of woodrat sign (Hassinger and Dunn 1990).

In line with this objective, investigators have developed operational definitions and a code manual (Butchkoski 2006) to enable teams to accomplish the visual surveys. Based on live-trapping experience, a *habitat site* is defined as an area of more-or-less contiguous surface rock, with no break in coverage greater than 200 m that contains one or more *activity centers*. Activity centers are defined as an area 30 m in diameter, that is centered on signs of fresh or old activity (toilet areas or midden caches) in surface rock (Hassinger et al. Chapter 8). The inhabitants of each habitat site were termed a subpopulation (Hunter 1996). In a visual assay, surveyors located habitat sites and then scored and recorded woodrat toilet areas and food caches. Toilet areas were categorized as fresh (intact, dark-colored, shiny) or old (disintegrated, gray, dull). Food caches also were categorized as fresh (intact, green, succulent) or old (disintegrated, faded, dried-out, fragile). In line with the experience from trapped sites, when fresh sign was found the habitat site was designated "active;" when only old sign was found the site was scored as "inactive."

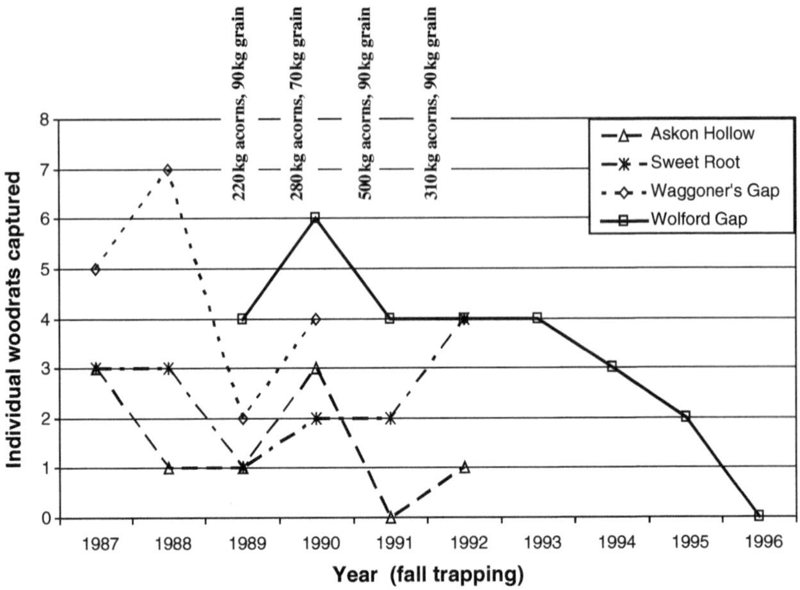

Fig. 7.1 Number of Allegheny woodrats (*Neotoma magister*) live-trapped at four Pennsylvania long-term monitoring sites: Askon Hollow Cave (Fayette County), Waggoner's Gap (Cumberland/Perry County), and Sweet Root and Wolford Gap (Bedford County). Food supplements for Wolford Gap site shown; supplements ended April 1993

Potential activity centers (suitable rocky habitat where no woodrat sign was found) were also recorded for future reference. Data collected from additional habitat sites included geographic location, habitat site size (length and width of rocky habitat), surface rock characteristics, topography, and vegetation. Data were then incorporated into a Microsoft® Access database and GIS (ESRI®/ArcMap, Environmental Systems Research Institute Inc.).

A second level of definitions was applied once the sites were mapped. Pennsylvania biologists defined a *metapopulation area* as an assemblage of habitat sites where migration from one subpopulation to at least some others is possible, at a distance of up to 2 km. The subpopulations seem to be able to exchange individuals and recolonize habitat sites that have recently become vacant. Operationally, topographically related habitat sites separated from the nearest metapopulation area by at least 10 km of habitat not suitable to woodrats or a significant barrier to dispersal (e.g., a river or farmed valley) were grouped together. Finally, a *conservation management area* (CMA) was defined as a set of physiographically related metapopulation areas that can feasibly be managed with a defined strategy. A CMA represents an economy of scale; different metapopulation areas within a CMA are likely to be impacted similarly by regional threats and public land management. Typically, in Pennsylvania, a CMA is a mountain complex or river drainage. Future management planning will be focused at the CMA level.

Trapping records from the long-term sites (Fig. 7.1) illustrate subpopulations winking out (Askon Hollow Cave) and numbers of individuals fluctuating from year to year. At Wolford Gap, the site with supplemental feeding, the population remained relatively stable, despite defoliation and tree mortality, until feeding ceased. Although the Wolford Gap site information is somewhat anecdotal, this is the only known case where a subpopulation received supplemental feeding, and the numbers suggested that there was a positive response. Sample size is too small to determine whether feeding decreased mortality or increased reproduction or immigration into the site. However, this case does point to the need for stable food resources on a site or, more importantly, within a metapopulation.

The program of visual site surveys (sometimes complemented by trapping) covered much of the state by 2000. Since that time, additional surveys have filled in details, and many sites have been resurveyed. The authors acknowledge that visual surveys cannot confirm the presence or absence of a woodrat at a site. Rather, the surveys indicate sites that may currently be occupied by woodrats or were historically occupied. From 1982 to 2006, Pennsylvania biologists have conducted 1,255 surveys at 802 habitat sites (including appropriate habitat where no evidence of current or previous occupation was found). Surveyors have found evidence of current or former woodrat occupation at 443 (55.2%) of these sites; 246 (55.5%) were scored as active, and 197 were categorized as no longer active.

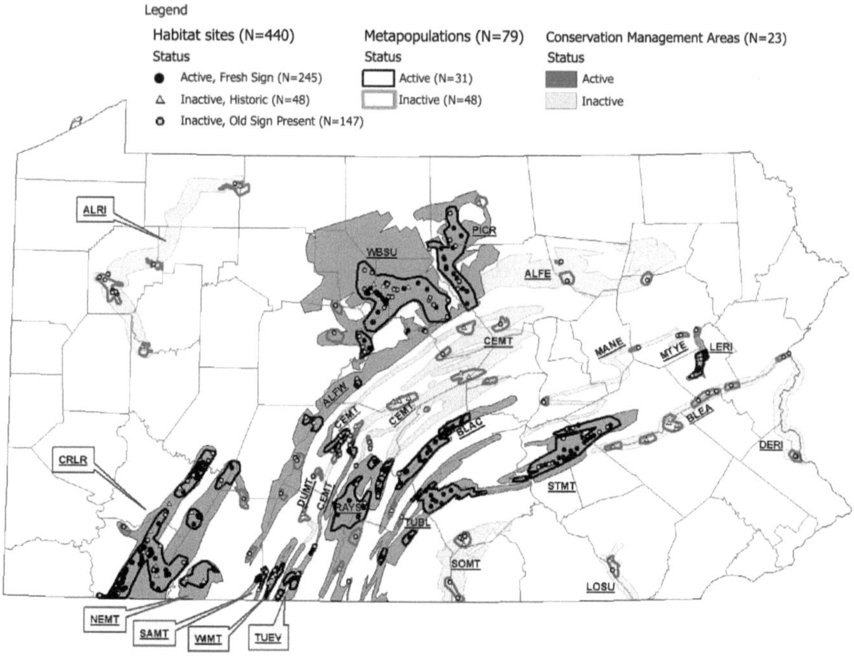

Fig. 7.2 Allegheny woodrat (*Neotoma magister*) surveys and landscape groupings used in Pennsylvania with labeled Conservation Management Areas (CMA)

Extirpation of woodrats from habitat sites, metapopulations, and CMAs all have been documented in Pennsylvania (Fig. 7.2). Although extirpation of a subpopulation is cause for concern, it is not uncommon for habitat sites to become vacant and then be recolonized while the metapopulation area (group of habitat sites) remains active. Thus, extirpation at a habitat site is not a reliable indicator of overall population decline. Therefore, in Pennsylvania the focus for analysis has shifted to documenting occupation at the metapopulation level (Fig. 7.2). We have no documentation of entire inactive metapopulations becoming recolonized. The same is true with the recolonization of CMAs, and 9 of the 23 CMAs are currently categorized as extirpated. The BLEA CMA (most eastern) is the most recent known extirpation, with the last habitat site becoming inactive during the fall/winter of 2003/2004 (Fig. 7.2). This CMA is being monitored continually. The ALRI CMA (most western) likely became extirpated between 1990 and 1991, given the large amount of old sign observed in April of 1992. Based on museum records and preserved sign in protected areas, the other seven CMAs were likely extirpated over 20 years ago. At least four CMAs (ALFW, CEMT, NEMT, and SAMT) have few and relatively small active metapopulations and may be considered threatened with extirpation (Fig. 7.2). Quite clearly, the Allegheny woodrat is in decline in Pennsylvania.

The incorporation of distribution data into GIS has allowed a rough model of the statewide distribution of woodrat habitat to be produced. Spatial data for the model were obtained from Pennsylvania's Spatial Data Access Website (http://www.pasda.psu.edu). Core habitat was modeled using land cover, slope, and surface geology found within 300 m of known active sites. This yielded a core habitat characterization consisting of a land cover of forest and slope of $>10^\circ$ on 20 surface geological formations in Pennsylvania (Butchkoski 2003, 2006). A map using this definition of appropriate habitat throughout the state demonstrates the linear, fragmented nature of the rare rocky habitat currently known to be available to woodrats in Pennsylvania (Fig. 7.3).

The modeled area (ha) of active habitat, with some reasonable assumptions about woodrat density, has allowed us to estimate the total number of Allegheny woodrats in Pennsylvania. Using the number of apparent occupied den areas documented on active-habitat polygons for each CMA (known sites/ha by CMA), and the density of five animals per 45 ha (high-density estimate) or 300 ha (low-density estimate) of active habitat resulted in a statewide population estimate of approximately 1,500–10,000 Allegheny woodrats. Currently, the estimate is a coarse one, but this figure is one that can be refined, and at the low end definitely emphasizes the rarity of the species on a landscape scale.

The visual survey technique used in Pennsylvania has been successful in satisfying our primary objective of inventorying the current range of woodrats in the state, given the large spatial distribution and rugged landscape this species uses. A bonus has been the identification of long-extirpated habitat sites where old sign has been found in protected locations, such as caves or deep rock crevices, with the caveat that with this method we are not able to say when the last inhabitants disappeared. A large number of data points also is valuable for identifying habitat characteristics, range trends, and connectivity or fragmentation that will be useful in managing the species on a landscape scale.

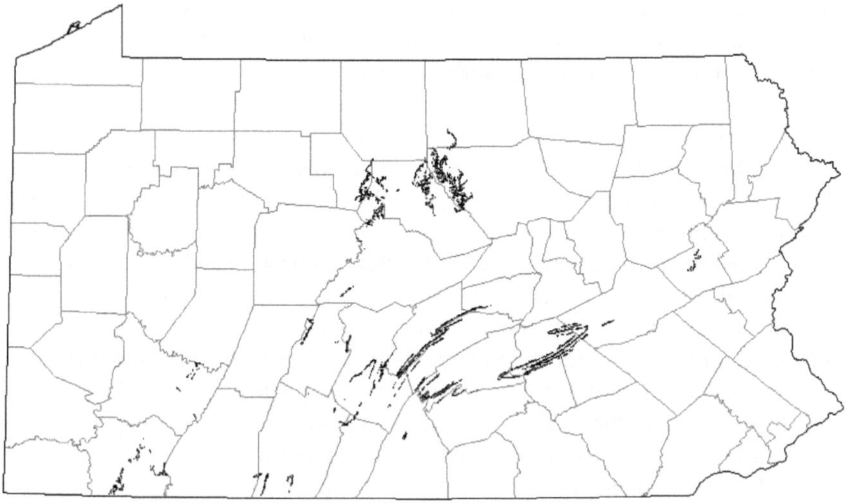

Fig. 7.3 Rocky habitat currently estimated in use by Allegheny woodrats (*Neotoma magister*) in Pennsylvania. Shading depicts the rocky habitat at or near currently known active woodrat habitat sites in Pennsylvania and was extracted from a statewide habitat model created using surface geology (rock type) used by woodrats in forest cover on a $>10^\circ$ slope. The figure also emphasizes the woodrat's dependence on geologic features (ridges and gorges) with a linear distribution

Coupled with the use of GIS, our surveys allow researchers to ask questions such as: How much fragmentation can a population tolerate on the habitat site, metapopulation, and CMA? What influence do food resources have on reproduction, dispersal, and recolonization? Do woodrats use landscape characteristics for dispersal? What are the potential barriers to dispersal? Our biggest information gaps at this point are data delineating microhabitat characteristics, behavior, and reproductive biology that would suggest mechanisms for large-scale patterns. These can be addressed in the future by the carefully designed manipulative studies at a more local scale.

Maryland Program

In 1990, as a response to reported population declines in northeastern states, Maryland's Department of Natural Resources Natural Heritage Program staff initiated an intensive survey to determine the status, distribution, and current population trends of the Allegheny woodrat. This effort included long-term population monitoring at six sites and intensive surveys of potential habitat across the state (Feller et al. 1994). For the long-term monitoring study, live-trapping methods were initially employed at four sites in 1990, and continued annually from July through September to minimize disruption of parturition and rearing of young. Biennial trapping was added at an additional two sites, with an overall term of 15–17 years for the six long-term sites. Sites encompassed a variety of habitats including limestone bluffs and quarries, ridge crest sandstone, and streamside metamorphic outcrops, selected to include at least one site in each of the four counties still occupied by woodrats.

Methods for live-trapping generally followed the protocol used by other states. Methods were modified slightly to conform to local habitat characteristics, but remained constant at each site. Ten to 35 ($\overline{X} = 24$) Tomahawk live traps baited with peanut butter and oats were placed (ca. 10–20 m apart) in protected overhangs, and adjacent to latrines or middens when possible. Marked trap locations at each site provided sampling consistency over time. Sex, mass, and reproductive condition was recorded for each capture prior to ear tattoo application and release.

A second effort was aimed at inventorying the distribution and abundance of woodrats throughout the state. Potential habitat in the form of caves, rock outcrops, cliffs, and large talus was identified throughout the state using topographic/geologic maps, leaf-off aerial photographs, museum records, publications, and knowledge of DNR field staff and others. The vast majority of ridge crests and steep areas in the woodrat's historic range were traversed by foot. In addition to revisiting all historic sites, field surveys documented woodrat sign at an additional 62 sites (Feller et al. 1994). A site was defined as a single or cluster of habitat islands occurring within a linear distance of ≤ 1 km and with no intervening barriers to dispersal, such as rivers, improved roads, clearings, or development (Lay and Baker 1938). Some of the visual survey sites ($n = 21$) were live-trapped at least once. In conjunction with the long-term sites, these trap sessions yielded 938 captures of 530 woodrats in 5114 trap-nights at 27 sites.

The vast amount of potential woodrat habitat identified from survey work, along with budget restrictions, required a more efficient alternative to live-trapping to determine site status and in estimating population size. A unique feature of the Maryland surveys was the development of quantitative latrine and midden surveys. These surveys permitted the assessment of presence/absence without the need for live-trapping. Fresh scats were defined by either a soft moist texture or black shiny exterior and were classified as old if dry with a dull, deteriorated exterior appearance. Based on visual observations, scat appearance transformed from fresh to old within a two-month period, the rate appearing to be influenced by temperature, humidity, and degree of exposure to weather. Fresh middens contained green vegetation, fresh cuttings, or recently acquired mast, while old middens had only dried vegetation, sticks, or empty nut hulls. Middens (or caches) constructed by woodrats have unique characteristics (Poole 1940), allowing them to be easily distinguished from those made by other animals.

A correlation of fresh latrine counts and live-trapping results conducted simultaneously at 13 localities ($n = 33$) over three seasons (spring, summer, and fall) showed fresh latrine counts in fall to be the best indicator of woodrat density ($r^2 = 0.80$, $n = 6$). Summer ($r^2 = 0.72$, $n = 14$) and spring ($r^2 = 0.59$, $n = 13$) indices were less reliable. Generally, an average of 2.1 fresh latrines was found for every woodrat captured. Very low woodrat densities at sites, particularly at shale barrens or other rock structures with limited deep crevice access, could give false negative results from scat surveys alone. This occurred at a proportion of sites during all three seasons and was most prevalent in summer (25%), and least in fall (17%). Presence of freshly cached vegetation in middens can reduce false negative determinations of current woodrat occupancy in the fall of the year. Using the quantitative scat and midden protocol, quantity and relative age (fresh or old) of woodrat latrines and middens were recorded during 265 intensive field surveys that investigated every accessible crevice, cave, and overhang.

At the six long-term sites, 3,478 trap-nights yielded 637 captures of 304 woodrats. Intensive trapping during the first year showed 25% of all marked animals absent from the original site within six months of initial capture. Rarely did a missing woodrat appear at an adjacent site or in a subsequent year, suggesting the absence was due to mortality. Long-term data showed a 37% recapture rate of woodrats during the second consecutive night of the two-night monitoring effort, while only 15% were recaptured the following year. Two male woodrats survived >5 years; this exceeds previously published reports for longevity in Allegheny woodrats (Mengak 2000, Mengak et al. 2002).

Lincoln-Peterson population estimates varied from 1 to 24 ($\overline{X} = 5$) woodrats/site, with peaks reached at all sites in the early to mid-1990s. Current (2006) population estimates of 1–6 ($\overline{X} = 1.5$) woodrats/site demonstrated the reduced abundance of woodrats at all monitoring sites. Each of the six monitoring sites was reduced to zero or one woodrat at least once since the late 1990s and was subsequently recolonized, presumably by immigrants from nearby sites, as is characteristic in a metapopulation.

The statewide survey (visual inspection plus trapping) identified 108 sites that woodrats have occupied at some time in Maryland. These included historic records, sites at which only old sign was found, and active sites with fresh sign. Between 1990 and 2006, only 50 (46%) of the 108 survey sites had current sign of woodrat activity (either fresh scats or fresh midden contents). This suggests a recent and severe woodrat population decline throughout Maryland. Since most of the sites were identified prior to 1993, declines were well underway prior to the start of the 1990 survey. Distribution of all 108 sites, historic and recently documented, shows a concentration in western counties (Fig. 7.4). Nearly 90% of Maryland woodrat den sites were within fractured sandstone or metamorphic rock that forms cliffs, smaller vertical outcrops, or large (>1 m) talus, though most frequently a combination of two or more rock forms occur at any one site. The remainders are shale barren outcrops, caves (Franz and Slifer 1971), mines, tunnels, and very rarely unoccupied buildings in remote settings. Den sites are integrated within an eastern deciduous forest community most frequently dominated or codominated by one or more oak species at present time. As evidenced by the presence of American chestnut (*Castanea dentata*) saplings at all Maryland woodrat sites, this species was likely an important canopy component prior to infection by blight (*Cryphonectria parasitica*). Statewide range remained similar to that as described earlier by Paradiso (1969), exceptions being first occurrence records in the Blue Ridge physiographic province and a failure to find any sign of woodrat habitation in the Piedmont (Fig. 7.4).

When results of live-trapping and periodically repeated latrine index surveys were combined, a reliable woodrat occupancy history over the course of the

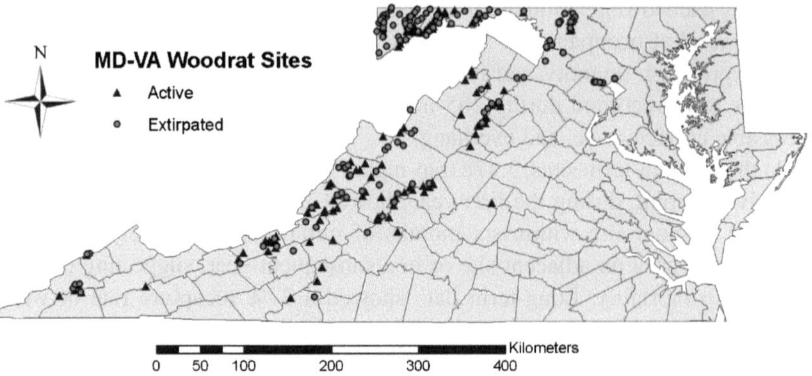

Fig. 7.4 Allegheny woodrat (*Neotoma magister*) survey sites in Maryland and Virginia. "Active" and "extirpated" designations refer to status of woodrat population in the most recent survey at the site since 1990

17-year study (1990–2006) was obtained for 62 of the 108 documented woodrat sites. These sites were termed "reference sites" because of high confidence in their status. Included were 229 periodic latrine index surveys ranging from 2 to 22 ($\overline{X} = 3.8$) surveys/site, conducted at 1–16 year intervals and bolstered by 134 live-trapping surveys at 27 of the 62 monitoring locations. Nineteen (31%) of these reference sites were scored as active for all or nearly all surveys, and 23 (37%) sites remained extirpated over the study period. The remaining 20 (32%) sites experienced status changes during the 17-year study. Two (3%) initially extirpated sites were subsequently recolonized, and 18 (29%) sites became extirpated and were not recolonized. At sites extirpated and not recolonized, increasing scat deterioration, as indicated by scat pellet surface deterioration, lower numbers of identifiable pellets within latrines, and ultimately fewer latrines were successively recorded.

Two case studies demonstrate the fate of extirpated sites, one with subsequent recolonization from the metapopulation matrix and the second occurring in an otherwise extirpated metapopulation. The first case study—demonstrating the extirpation of a site with subsequent recolonization—occurred at High Rock (Garrett County). With a complete seasonal to annual mark-recapture population history since 1990, this site was known to support the largest woodrat population in Maryland. Within the Savage River State Forest "wildland," High Rock is actually part of a larger site/metapopulation complex (Big Savage Mt. South) consisting of 3 km of discontinuous sandstone cliff/boulder ridge crest habitat at nearly 910 m elevation and several smaller scattered outcrops on the adjacent slopes. None of the 13 subsites comprising Big Savage Mt. South, including High Rock, are separated by more than 1 km, but most are >500 m apart. Mature red and white oaks, including Maryland's largest tract of old growth forest, dominate the area.

High Rock consistently supported 10–24 woodrats (based on 70 trap-nights of effort) until July 1998, when the population was found to be extirpated following a record peak population in 1997. Though all subsites of the Big Savage metapopulation had supported active woodrat populations in all previous surveys, a scat/midden survey in 1998 at nine subsites failed to find any recent sign of woodrat habitation. In fact, the most remote and extensive subsite, Heartline Heights, had 50 old latrines and 15 middens without any fresh material in September 1998. While the actual cause of woodrat decline is not definitively known, a complete failure of oak mast production was noted in the region during the fall of 1997 and may have contributed to the decline. Subsequently, in July 1999, two males were captured at High Rock, confirming that woodrats had survived or were undetected at scat/midden survey sites or at one of the four small unsurveyed subsites. One male survived to 2000, and with a single immigrant female, successfully produced two offspring in 2001. Woodrat numbers increased to 14 by 2002, possibly supplemented by additional immigration, until a devastating ice storm in the fall of 2003 killed or severely damaged virtually all large oaks on or near the ridge crest. From 2003 to 2006 only one or two male woodrats have been captured at High Rock. South slope and off crest sites in the metapopulation have recovered to a greater degree as evidenced by rebounding scat and midden indices. Recovery of High Rock and other ridge crest subsites is expected to be slow and woodrats may not reach previous densities for several decades if oak mast production is the limiting factor. Under adverse climatic conditions and resulting habitat degradation over a significant proportion of Big Savage South, Allegheny woodrats have managed to survive as a metapopulation within the protected forest and rock outcrop matrix.

The second case study—demonstrating the final extirpation in a metapopulation—occurred at Vansickles Rock (Garrett County). Vansickles Rock, a 300-m long mid-slope sandstone cliff above the Youghiogheny River also was first live-trapped in 1990. Vansickles Rock supported a dense woodrat population with 19 (spring), 11 (summer), and 8 (fall) captures (based on 50 trap-nights of effort), reflecting a decline through the seasons, and a pattern of decline also observed at High Rock and three other sites that year. Live-trapping in spring 1991 resulted in no captures, marking the loss of the last subpopulation of woodrats in the Youghiogheny River valley metapopulation, an area that once supported nearly half of all known Maryland woodrat subpopulations. Situated in a narrow protected band of forest, cliffs and smaller outcrops are scattered along an 18 km stretch on both sides of the high gradient river. Although a state designation of "wild and scenic" implies a degree of wilderness and ecological integrity, the narrow corridor is bordered by agricultural fields, reclaimed strip mines, and rapidly expanding residential development. Poorly regulated timber harvesting also occurs within the "wild and scenic" boundary. Isolated from West Virginia sites to the west, Pennsylvania sites to the north, and other Maryland sites to the south and east by residential development, numerous roads, and large agricultural fields, this metapopulation area has not been repopulated since its extirpation in 1991, and future recolonization is unlikely.

Following extirpation of woodrats over the winter or early spring of 1990–1991 at Vansickles Rock, latrines and midden index surveys recorded the rate of degradation over a period of 11 years. The initial survey was conducted while the site supported a population of eight woodrats during the fall of 1990, with five subsequent surveys 1–5 years apart. Latrines decayed linearly with 30% of original latrines identifiable as woodrat sign 11 years later. Uneven deterioration of middens was partially attributable to cache size and composition where woody items and nut shells greatly extended the period of detectability. Caches of only herbaceous vegetation were quickly lost in the first two years. Direct exposure to weather deteriorated latrines and middens rapidly, with decade-old sign persisting only within the deepest dry crevices and overhangs.

The downward trend of the Allegheny woodrat in Maryland is evident in survey and long-term monitoring data. Only 33 of 108 documented woodrat sites were found to be active when last monitored between 1990 and 2006. A 26% net loss of 62 reference sites over the 17-year study period indicates an average rate of site extirpation of 1.05% per year. Woodrat density at long-term monitoring sites dramatically declined. Recent capture rates are less than 33% of long-term averages and peak populations at most sites occurred over 10 years ago. The extirpation of woodrat subpopulations in the Youghiogheny River Valley in western Maryland and others in the Potomac Gorge area through the northernmost extent of the southern Blue Ridge physiographic province to the east has resulted in a longitudinal range contraction in the state (Fig. 7.4). Occupied sites are concentrated in the two most heavily forested areas of the state, Allegany County and eastern Garrett County. These areas will be targeted for conservation, though long-term viability is not certain.

We did not formally analyze possible causative factors for the decline, but the large amount of intensive field work gives us basis for comment on previously proposed mechanisms. Raccoon roundworm (*Baylisascaris procyonis*), a known cause of woodrat mortality (McGowan 1993), was present in 17.3% of raccoon feces collected at Maryland woodrat sites. Roundworms were most prevalent in both the Blue Ridge Province and Youghiogheny River Valley, areas suffering greatest woodrat decline. A raccoon fecal sample collected at Vansickles Rock in 1993 tested positive for raccoon roundworm, though five samples collected in the late 1990s were negative. However, behavioral observations of 983 captured woodrats and four brain necropsies of deceased individuals failed to detect raccoon roundworm infection. Additional research in woodrat mortality attributable to raccoon roundworm is needed in areas of Maryland and surrounding states.

Failure of oak mast is a suspected cause of woodrat decline, and possibly only a critical factor since the functional loss of the American chestnut (LoGiudice Chapter 2). In Maryland, American chestnut saplings were observed at all woodrat sites and cached nuts were found on several occasions. At Vansickles Rock, gypsy moth infestation was exceptionally heavy in 1990, which eventually led to a varying degree of localized oak mortality. Landscape-level fragmentation has been implicated as both a direct and indirect cause of woodrat decline (Balcom and Yahner 1996, Ford et al. 2006), though detection through landscape analyses has been difficult. The greatest current threat to habitat is the development of ridge crests and upper slopes by large-scale wind power projects on both private and public land. To obtain scenic vistas, recent residential developments also have focused on these areas, fragmenting the last remaining large contiguous tracts of privately owned forest. More detailed information regarding fragmentation effects on woodrat population viability is needed to reduce or eliminate impacts from such projects.

In Maryland, comprehensive survey coverage was possible due to the small geographic area which woodrats inhabit within the state. The rate of decline found in Maryland would not have been detectable with less intensive sampling. Other states with greater areas and less monitoring effort could be experiencing similar declines that are going undetected. With time and budget restraints, the effort in Maryland has been somewhat opportunistic and less systematic than ideal. We recognize a need for more complete investigation of woodrat dispersal behavior, potential barriers, and metapopulation viability to provide natural resource managers the tools to make better decisions during environmental review processes. Indices of annual mast production from long-term mark-recapture monitoring sites could provide insight to food related population dynamics. Annual monitoring of *B. procyonis* infection rates and density of raccoon feces at long-term monitoring sites would determine temporal trends and relative contribution to woodrat mortality.

Indiana Program

Indiana lies at the northwestern fringe of the current geographic distribution of the Allegheny woodrat in the eastern United States (Hickie and Harrison 1930, Wright Chapter 1). Although woodrat fossils are fairly widespread in caves in Indiana (Richards 1987), the Allegheny woodrat currently has a greatly contracted range limited to limestone escarpments bordering the Ohio River in Crawford and Harrison Counties in extreme south-central Indiana. Accordingly, the Indiana Department of Natural Resources (IDNR) listed the Allegheny woodrat as a state-endangered species in 1984 due to its reduced geographic distribution.

In 1991, the IDNR initiated a live-trapping program to monitor the status, distribution, and relative abundance of Allegheny woodrats in Indiana. Initial trapping sites were those reported by Cudmore (1985) to have extant woodrat colonies, which included 18 bluffs and two caves distributed along a 74-km stretch of the Ohio River from Evans Landing in Harrison County downstream to Alton in Crawford County. These sites correspond to the designation "historic sites" used by other states and outlined in this chapter. The original site boundaries delineated by Cudmore (1983) were retained, although woodrats move among adjacent sites that encompass one contiguous cliff line (Johnson 2002). In such situations, individuals recaptured at a second site in the same survey were assigned to their initial capture site. Three previously undocumented sites (Noes Park Bluff, P.P. Pot Cave, Limekiln Hollow Cave) reported by Johnson (2002) were incorporated into the monitoring program after their initial discoveries in 1992 and 1993.

As the distribution of woodrats in Indiana is so limited, we did not distinguish between long-term monitoring sites and snapshot survey sites as was done in Pennsylvania, Maryland, and Virginia, but instead had live-trapped at all sites (except three bluff sites for which landowner permission was denied after the first survey) periodically from 1991 to 2006. Trapping was generally conducted from mid-June through October to reduce possible disturbances during peak parturition and pup-rearing periods (i.e., April and May; Cudmore 1983). This schedule was a modification of the 40-trap, two-night protocol used by other states. At bluff sites, 22–65 ($\overline{X} = 44.2$) traps were set yielding a mean of 88.2 (± 24.1 SD) trap-nights per session. Trap effort at caves was less ($\overline{X} = 23.8 \pm 22.0$ trap-nights/session) and depended on passage length, complexity, and presence of rock outcrops near the entrance. Sex, age class, and mass (to nearest 2.5 g) were noted for each woodrat captured. Age class was estimated from mass at initial capture (i.e., juvenile <175 g; subadult 175–224 g; adult ≥225 g), reproductive condition, and pelage coloration (Mengak 2002a).

Six range-wide surveys, comprising 93 trapping sessions at 21 sites (17 bluffs, 4 caves), were conducted in Indiana between October 1991 and September 2006. At least one woodrat was captured during 73 (78%) sessions. Elapsed time between surveys at respective sites ranged from 1 to 6 years ($\overline{X} = 3.0$). Mean length of bluff surveyed was 859 ± 386 m. A total of 531 woodrats was captured 960 times in 7106 trap-nights at 15 (71%) sites (Fig. 7.5). Distribution by sex/age class at initial capture of 523 individuals was 143 (27%) adult males, 165 (32%) adult females, 46 (9%) subadult males, 85 (16%) subadult females, 48 (9%) juvenile males, and 36 (7%) juvenile females. Overall capture rate corrected for sprung traps was 14.3 woodrats/100 trap-nights. Relative abundance at occupied bluff sites ranged from 1.2 to 35.0 woodrats/km of bluff ($\overline{X} = 11.9 \pm 7.9$). These are minimal values because it is unlikely all residents were captured in only two nights, and estimates may be less at sites surveyed before juveniles become independent.

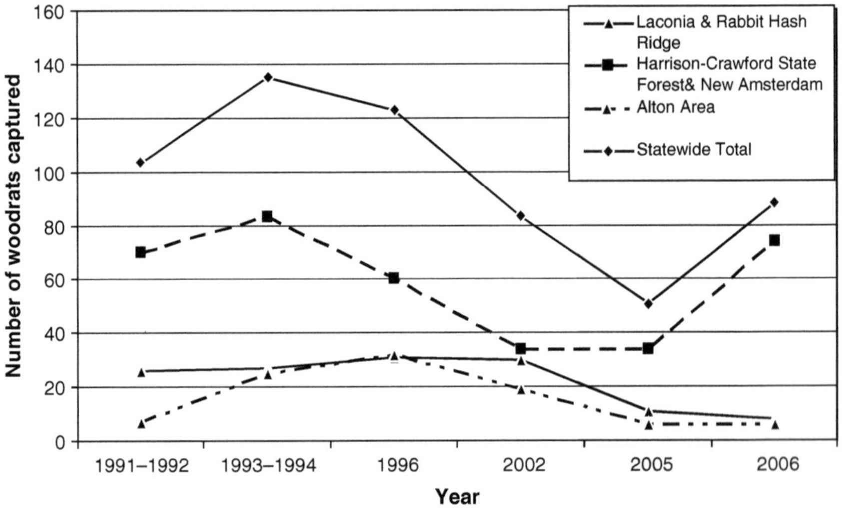

Fig. 7.5 Combined live-trapping results at Allegheny woodrat (*Neotoma magister*) long-term monitoring sites in Crawford County and Harrison County, Indiana. Number of individual sites in each area: Laconia & Rabbit Hash—7 bluff sites; Harrison-Crawford State Forest & New Amsterdam—6 bluff, 4 cave sites; Alton—1 bluff site. Four bluff sites in Leavenworth & Fredonia area were trapped in 1991–1992 with no woodrats captured; no subsequent trapping at this site (data not shown)

An average of 8.8 (± 8.0) woodrats were captured at occupied sites. A single woodrat was caught in each of 14 trapping sessions at six sites including three small caves approximately 1 km from the Ohio River. In contrast, four bluff sites (Bull's Point Bluff, Cold Friday Hollow Bluff, Shelterhouse #2 Bluff, South HCSF Bluff) each yielded ≥15 woodrats in 14 of 23 (61%) trapping sessions. Most ($n = 351$; 66%) of the woodrats captured in Indiana originated from these four sites (Fig. 7.5).

We observed both long-term occupancy and local extirpations. Seven sites produced ≥1 woodrat each time they were trapped ($n = 42$). Monitoring efforts to date have suggested local extirpations of woodrat colonies in Indiana at 13 of 20 sites (65%) that were occupied in the 1980s. The initial range-wide survey in 1991–1992 found no woodrats at five sites that had harbored woodrats >10 years earlier, including Horseshoe Bend Bluff, which had yielded 22 individuals (Cudmore 1983). Although suitable rock structure remained, these sites lacked any fresh woodrat sign and no animals were captured in 400 total trap-nights. Woodrats were apparently extirpated from eight additional sites during our survey period.

Twelve of the 13 sites, from which woodrats were believed to be extirpated, were revisited one or two times each to assess the species' capacity to recolonize formerly occupied habitats. Visits consisted of both site inspections for fresh sign and live-trapping sessions. Several winter bat censuses were conducted in one cave, thus providing ample opportunities to document recolonization. Only two cases of recolonization of formerly occupied sites were confirmed in Indiana. In September 2006, 15 woodrats (4 males, 11 females) were captured at Shelterhouse #2 Bluff, approximately five years after the population was believed to have been extirpated. Similarly, a young adult male recolonized Potato Run Cave between August 2005 and September 2006. This small cave has yielded only a single animal in four previous surveys and was last known to be occupied by an adult female nearly four years previously.

Multiple lines of evidence from monitoring efforts over the last 15 years suggest a steady decline in Indiana's remnant woodrat population. In 1993–1994, woodrats occupied 15 sites with a cumulative total of 135 individuals statewide. The following three surveys spanning 11 years, however, showed local extirpations coupled with rapidly diminishing relative abundance at active sites. By 2005, only 51 woodrats were captured at 8 of 13 sites. Surveys in 2006 documented a 73% increase in number of woodrats captured, but this apparent increase may have been partly artifact, because trapping was conducted shortly after peak parturition and pup-rearing periods.

Although the limits to the species' geographic range in Indiana have remained largely unchanged for >25 years, the distribution of Allegheny woodrats in Indiana is characterized by small, spatially isolated populations separated by extended stretches of less suitable, unoccupied bluffs. Vacant habitats are often typified by marginal rock features that lack deep fissures, ledges, and crevices needed for protection, food storage, and denning opportunities. In 2006, extant colonies occupied approximately 7 km of linear cliff face, less than 10% of the species' recognized range limits in Indiana.

Indiana has been unique among states with active monitoring programs wherein most, if not all, known extant sites can be effectively surveyed within a single field season. Allegheny woodrats are habitat specialists with strong affinities for cliffs, outcrops, talus slopes, caves, and other rocky habitats in deciduous forested communities. In Indiana, such conditions are limited to linear segments of wooded bluff habitats in portions of only two counties. Consequently, live-trapping surveys may generate a more complete population estimate than in other regions of the species' range where suitable habitats are widespread and perhaps less accessible.

Because Indiana's monitoring program has employed a consistent protocol since its inception, additional data beyond presence and relative abundance may be gleaned from field surveys. For example, movement patterns, reproductive parameters, longevity, recapture rates, and simple estimates of survival are possible from existing data. Optimal microhabitat attributes sought in den site selection can be attained where particular crevices or fissures were consistently occupied by different individuals. What is lacking from Indiana are concurrent data for independent variables that may explain trends in occupancy and population indices observed during long-term monitoring programs. Habitat quality (e.g., species composition, extent of fragmentation), food habits and food availability (e.g., cache contents, annual hard mast production), incidence of *B. procyonis* in the environment (e.g., raccoon latrine densities, infection rates), and dispersal behavior are examples of information needed to better understand the mechanisms that influence Allegheny woodrats in Indiana.

Virginia Program

Relatively little historical information exists for Allegheny woodrats in the southern Appalachian region. Bailey (1946) and Handley and Patton (1947) listed Virginia counties known to have woodrats, but more recent distribution had not been studied before 1990. Beginning in 1989, we combined literature records of *N. magister* with museum records for Virginia from the National Museum of Natural History (Smithsonian Institution). We also contacted cavers from the VPI Cave Club in Blacksburg, Virginia and obtained an informal list from a researcher at the Virginia Commonwealth University (J. Pagels pers. comm.). These museum records, field observations, and anecdotal reports formed the basis of a list of known woodrat observations in Virginia. The sites on this list were designated as "historic sites" if the observations were made prior to 1985 (i.e., five years prior to the beginning of this study). Records from 1985 to 1990 were designated recent sites.

From the database, we plotted reported locations of *N. magister* in Virginia on topographic maps. The sites were then visited and live-trapped to ascertain their current occupancy status. Sites with inadequate location data were not visited. Forest Service, National Park Service, and Virginia Department of Game and Inland Fisheries personnel suggested additional sites. These sites were surveyed as funds became available during the course of our 12 years of surveys. Personnel of these agencies assisted in surveying for woodrats when possible.

Most sites were trapped only once throughout the study to document the statewide distribution and occurrence (i.e., presence or absence) of woodrats, primarily on the national forest. Nine sites (eight on national forest and one in Shenandoah National Park) were trapped approximately annually in the fall, September–October (Mengak 2002b). These nine sites were designated as LTM sites. Sites that were trapped only once or revisited infrequently were designated survey sites. Two LTM sites (Bath County and Giles County) were trapped at approximately 60-day intervals (weather permitting) from April 1996 to May 2000. All sites were trapped and treated identically; only the frequency of trapping differed between sites. Selection of LTM sites was primarily based on accessibility, known current occurrence of woodrats, historic occurrence of woodrats, and extent of the habitat and population.

Trapping effort generally involved setting 10–15 traps for one or two nights at nearly all survey sites. Occasionally, as few as five traps were set at a survey site if the area was too small to accommodate a greater number. At first, traps were set for only one night, but later a second night of trapping was added if no animals were caught on the first night. At the LTM sites trapped annually, including the two sites trapped bimonthly, 40 traps were set for two nights. Occasionally, an LTM site was trapped for three or four nights in order to validate the sampling procedure, but only the first two nights of trapping were used in the analysis of population trends.

Sites were classified on the basis of trapping results as active (at least one woodrat captured) and inactive (site trapped but no woodrats captured). All captured animals from both survey sites and LTM sites were treated identically. Captured animals were ear-tagged in both ears, weighed, sexed, examined for reproductive condition, ectoparasites, and overall body condition, and released at the point of capture. Captured animals were aged based on a combination of weight (juveniles ≤ 175 g, subadults 176–225 g, adults ≥ 225 g), pelage color, and reproductive condition (palpation of fetuses; evidence of lactation—bare, exposed teats or expression of milk; scrotal testes).

Fieldwork at LTM sites began in October 1990 and ended in October 2001. During the study, some LTM sites were discontinued from trapping and other LTM sites were added, such that not all nine LTM sites were used in all 12 years of the study. Once an LTM site was dropped from the study it was not trapped at a later date. Sites were dropped or added based on shifting budget priorities, available personnel, or site-specific objectives. Two of the nine LTM sites were trapped for the entire 12 years.

Survey sites were trapped between June 1990 and August 2000. Trapping effort at survey sites was not distributed evenly throughout the year, but some trapping occurred in every month. Most trapping occurred in the spring (May–June) or fall (September–October). A total of 189 sites were trapped; 22 (15.3%) were trapped more than once to examine whether woodrat sites remain occupied or if the site is unoccupied for a time.

We used several approaches to analyze the population size at LTM sites and persistence at trapping sites. First, a catch-per-effort index (ratio of captures to number of trap-nights corrected for sprung traps) was calculated for each monitoring site. Second, data from two nights of live-trapping events were used to compute

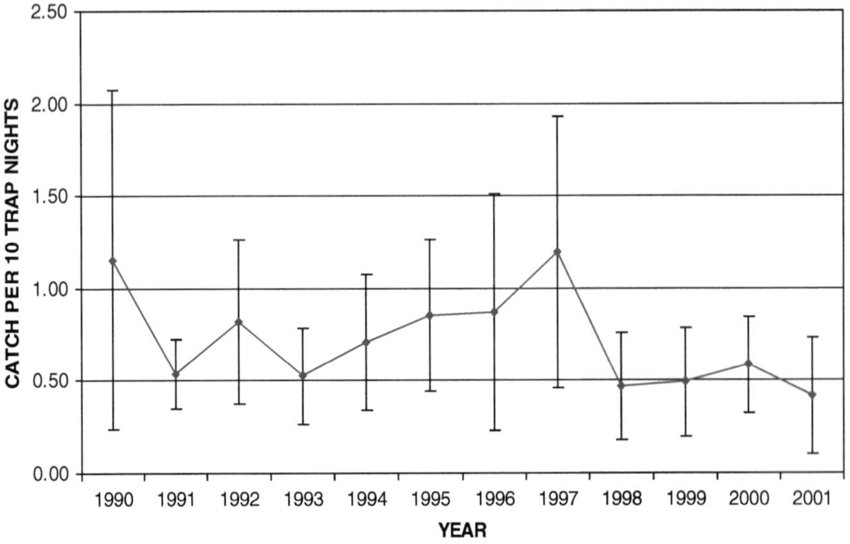

Fig. 7.6 Mean (±2SE) number of woodrats (*Neotoma magister*) caught at long-term monitoring (LTM) sites in Virginia from 1990 to 2001. The number of sites trapped each year varied from four sites to nine sites

the Lincoln-Peterson estimate of population size at each LTM monitoring site for each trap session (Thompson et al. 1998). The expanse of the monitoring site (ha) was not determined, so no estimates of density are available. Additional population modeling and analysis was conducted for LTM sites, where 10 or more woodrats were captured during the annual fall monitoring. Closed population models from Program CAPTURE (http://www.mbr-pwrc.usgs.gov/software.html#a; Otis et al. 1978, White et al. 1982) were used to estimate population size, confidence intervals, and probability of capture. At all the nine LTM sites, the annual number of captures fluctuated, often moving from highest to lowest in one to two years. The total number of woodrats captured at an LTM site during the annual fall monitoring varied from a high of 20 individuals to a low of zero (Fig. 7.6).

The actual number of captures and recaptures was used to calculate three population estimators at the long-term monitoring sites. Two of these, the Lincoln-Peterson index and the catch-per-effort index, were closely correlated (r^2 = 0.987 for Butt Mountain; 0.945 for Star Chapel; 0.947 for Rocky Mountain; 0.991 for Devil's Marbleyard; 0.938 for Elizabeth Furnace; 0.943 for Bluff's Trail; 0.705 for The Stacks; and 0.462 for Sulphur Springs Gap; Fig. 7.7). These values suggest that the monitoring protocol was very successful in catching the majority of the woodrats present at the site.

The third population estimator, an estimate and 95% confidence interval, was appropriate when the absolute number of captures was ≥10 individuals. Seven LTM observations fit the criteria. Our analysis revealed that the estimated woodrat population size varied from 15 to 20 individuals again agreeing closely with number of

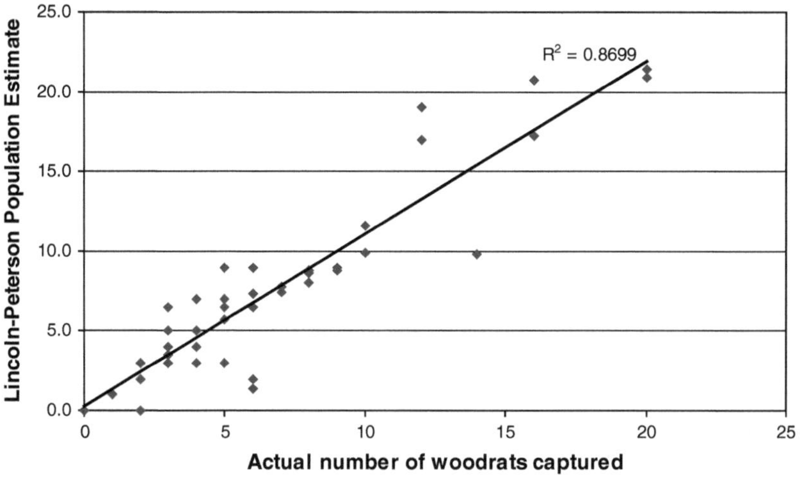

Fig. 7.7 Relationship between numbers of Allegheny woodrats (*Neotoma magister*) caught at long-term monitoring (LTM) sites in Virginia, 1990–2001, and Lincoln-Peterson population estimate for the site based on analysis of mark-recapture data. The number of sites monitored each year varied from four sites to nine sites. All trapping was conducted between mid-September and mid-October in each year

actual captures. The 95% confidence intervals ranged from 11 to 33 individuals. Only at Butt Mountain in 1990 did the actual catch exceed the population estimate. Confidence interval estimates were generally narrow except at The Stacks in 1997 and Star Chapel Cave in 2000. In both cases, 12 individual woodrats were caught in two nights of trapping but only four and three woodrats were recaptured on the second night at The Stacks and Star Chapel Cave, respectively. This resulted in the high standard error estimates and wide confidence intervals.

At an individual LTM site, populations fluctuated widely from year to year, but there appeared to be no consistency from site to site in the timing of high or low capture indexes. Woodrat populations showed a declining trend at Butt Mountain, Devil's Marbleyard, and The Stacks. The trend was increasing at Star Chapel Cave largely on the strength of the 2000 sample year. The overall trend in abundance was flat or declining at the other LTM sites (Mengak 2002b). The regression coefficient (r^2 = 0.67) was significant (p = 0.012) only at Devils Marbleyard. The sites are scattered across the state; Butt Mountain, The Stacks, and Star Chapel Cave are in the Valley and Ridge Province while Devil's Marbleyard is in the Blue Ridge.

Two LTM sites were trapped at times other than the annual monitoring periods. Butt Mountain, Giles County was trapped 24 times over 11 years including 3–4 times per year from 1996 to 1999. The Butt Mountain population fluctuated from a high population in 1990 to a low in 1991 followed by another high in 1996 through 1997 and a total collapse in 1998 (Mengak 2002b). Of particular interest is the period 1996–1998. During the period, the population was at its highest level for two years and completely disappeared in a span of six months. In 1997, 32 different individuals were regularly trapped at this site. Six months later (April 1998) no woodrats were captured. This situation persisted for three trap sessions in 1998 until a lone, unmarked male was captured in October 1998. The population began a slow recovery to five individuals in May 2000 when all trapping ceased.

The second intensively trapped site was Star Chapel Cave in Bath County. This site was trapped 30 times in 11 years including monthly in fall 1993 and every other month for most of 1998 and 1999 (weather permitting). The population size was generally low (i.e., 2–6 animals) but increased to 15 animals in June 1999 (Mengak 2002b). The fluctuation and collapse seen at Butt Mountain was not seen at Star Chapel.

The data from survey sites added the spatial dimension of woodrat status in Virginia. Survey sites were classified as woodrats present or absent. Presence–absence surveys were analyzed with program PRESENCE (http://www.mbr-pwrc.usgs.gov/software.html#a). This program estimates patch occupancy rates and related parameters including an estimate of the probability of detecting a woodrat at a site (MacKenzie et al. 2006). Output from the simulations was used to estimate the number of sites needed to detect various levels in change in mean occupancy rates (Parkinson et al. 1988) for future surveys.

In this study, we identified 253 sites as potential woodrat habitat. Forty-four (17.4%) were historic sites with records dating from as early as 1896 (Potomac River, 16 km above Washington, D.C.; USNM Number 086450). The rest (82.6%) were added during the study. Of the 44 historic sites, 14 (31.8%) were active at the time of trapping. Of the recently added sites, 105 (50.2%) were active at the time of trapping or observation (Mengak 2002b). Therefore, trapping data or reliable observations (from mammalogists, USFS, or DGIF biologists) are available for 191 (75.5%) of the 253 sites identified in this study. Of these sites, 119 (62.3%) were occupied by at least one Allegheny woodrat on the date of trapping or observation (Mengak 2002b).

Of the 145 survey sites that we trapped in this study, 22 (15.3%) were trapped more than once. This allowed us to examine whether woodrat sites remain occupied or if the site is unoccupied for a time—"local extinction" in the terminology of conservation biologists (Meffe and Carroll ¨1997). Nine survey sites were trapped twice. Of these nine sites, one (Chimney Run) was not occupied on either trapping occasion and four were occupied on only one occasion. Of the 10 sites trapped three times, only six were occupied on all three occasions. Of three sites trapped four times, one was occupied twice and two were occupied three of the four times they were trapped (Mengak 2002b).

We used survey site data and the number of individuals captured at each site in Program PRESENCE to estimate a detection probability and to determine the sample size needed to detect a change in site occupancy rates. The average simulation-based estimate of site occupancy was 85.5% compared to the observed site occupancy of 64.1% (Mengak 2002b). The minimum sample size needed to observe a change in the number of sites occupied would be 27 sites in order to detect a 10% change in mean occupancy rates with the error rates set at 0.05 and 0.1 for Type I and Type II errors, respectively. Changing the error rates and the minimum detectable change in mean occupancy levels increases or decreases the sample size (Mengak 2002b).

One clear pattern from both the long-term and survey data is that local populations of Allegheny woodrats are small. Four measures of population abundance were available—actual number of woodrats caught, catch per 10 trap-nights, Lincoln-Peterson estimator, and CAPTURE output. As expected, results from each of these measures were in close agreement. Further, each measure revealed that the overall abundance of woodrats at an LTM site was low (i.e., captures varied between 0 and 20).

While the geographically large distribution of woodrats across Virginia is not in doubt, their long-term survival prospects are not certain. They occur in isolated locations with few individuals per location and commonly disappear from individual locations. Even sites with relative large populations (i.e., several LTM sites) declined to zero within a short time. Low abundance of woodrats at LTM sites (presumably the most extensive habitat available, hence their inclusion as an LTM site) along with the geographically separated nature of survey sites which often contained only one or two individuals, clearly indicates the sparse nature of woodrat distribution. It appears that woodrats in the Ridge and Valley and Blue Ridge Provinces of Virginia are functioning as isolated subpopulations and the entire woodrat complex is acting as a metapopulation.

Conservation and Management of Declining Species

The Allegheny woodrat monitoring and surveying programs in these four model states indicate a range of approaches and emphases in the monitoring of this species. Pennsylvania's program has put most effort into developing a standardized visual survey protocol that can be used, in conjunction with GIS mapping habitat data, to analyze spatial factors responsible for the decline of the species. The definitions and protocols formalized into a field manual in Pennsylvania (Butchkoski 2006) can be incorporated into training workshops for foresters, biologists, and other personnel, and can be used throughout the woodrat's range for standardized data collection. The GIS database has been important in several ways. It has allowed a continually improving delineation of woodrat metapopulation complexes in Pennsylvania and more informed planning for what areas to survey or resurvey in each field season. GIS also makes possible timely communication with agencies involved in environmental impact assessment. To some degree, GIS has allowed testing of hypotheses about habitat quality, mast supply, and distribution of habitat sites in relation to population status, and in aiding the planning of management sites and buffer zones (Hassinger et al. Chapter 8). To achieve its survey goals, Pennsylvania curtailed its long-term monitoring program, but data from repeat surveys of some sites have demonstrated extirpations and recolonizations as have been observed elsewhere.

The program sponsored by Maryland, an intensive combination of live-trapping and visual surveys, gives a comprehensive picture of woodrat populations in a relatively small state. The strength of the Maryland program has been its emphasis on developing interpretable data from visual observation. The time-course studies of breakdown of food-cache middens and latrines allow for more accurate scoring of extirpated sites, and the calibration of latrine counts with trapping results enables a semiquantitative population estimate from visual surveys.

The intensive census effort sponsored by Indiana on its few remaining woodrat sites amounts to a virtual annual enumeration of Allegheny woodrats in that state, along with detailed location information. There is valuable potential in these data for comparing the details of demography of successful sites with declining ones, or for defining the likelihood of recovery after a down year. There are also instances of individuals dispersing among sites that may give insight into the conditions that lead to recolonization.

Monitoring and survey studies in Pennsylvania, Maryland, and Indiana were initiated after the Allegheny woodrat had already been designated a threatened (Pennsylvania) or endangered (Maryland and Indiana) species. Their potential for contrasting stable or improving populations with declining populations is therefore limited. Virginia is in a special position, in that *N. magister* has not been designated a species of special concern in that state. Some of the long-term data from the Virginia program support the idea that the species is secure there, but data from some sites indicate decline, and the statewide surveys show an ominous trend of disappearance particularly from northern counties. Virginia data demonstrate the importance of understanding demography and natural history in healthy populations in order to interpret decline and disappearance. The Virginia program's emphasis on statistical modeling and interpretation helps to detect which differences are real and which are insignificant fluctuations. Taken together, studies in these four states point out that although live-trapping identifies the majority of animals, it is still a sample rather than a census, and some sites that appear to be empty of woodrats may still be occupied. Repeated samples are necessary to track recolonization or to give managers confidence in designating a site as inactive or unoccupied.

The question of live-trapping versus less labor-intensive (and potentially disruptive) assessment methods has come up in all four programs. Virginia and Indiana have strongly maintained the perspective that only live-trapping (despite the uncertainty in declaring a site unoccupied if no individuals are captured) can give a truly definitive indication of woodrat occurrence. Maryland and Pennsylvania have implemented nontrapping survey techniques with the conviction that the resulting data, with suitable caveats, can answer carefully defined questions. To meld the two approaches, calibration is crucial. For example, live-trapping at a set of sites visually classed as active could provide valuable information about the validity of the Pennsylvania surveys. Similar activities in Maryland help increase the confidence in visual survey data.

All four states have made adjustments in their trapping and survey protocol as work has progressed, and all admit shortcomings. None has consistently quantified the possible causative factors (forest cover, raccoon roundworm presence, food availability, etc.) that have been hypothesized to account for woodrat population declines, with the result that tests of the hypotheses have been limited with these data. We recommend that collection of systematic, quantitative data on such factors be a part of the planning for long-term or survey efforts. Sites for long-term studies tended to some extent to be sites of convenience, or sites chosen for probable persistence through time. We recommend that future long-term studies should be planned carefully to include designs that will yield a maximum of information, including some sites which are under threat and sites (like Wolford Gap in Pennsylvania) for which special management efforts are being made.

Although any program for conservation of a declining species must be tailored to the region and resources available, we urge communication and cooperation among agencies and researchers to maximize the usefulness of the data gathered. In the programs described in this chapter, differences in terminology, protocols, and analysis have resulted in data incomparability at times among the four states, even within a research network that is relatively connected. Workshops, manuals, and database planning take time, but can pay off in the long run in reducing wasted effort and making results more available.

In summary, lessons learned from long-term studies of the Allegheny woodrat could not have been learned with shorter duration projects. The issue of sign surveys versus actually capturing an animal continues to cause debate among investigators. The need to balance data requirements with the harsh realities of limited time and monetary resources is an important factor in deciding how to structure a monitoring program. The lesson from Virginia is that when effort is expended on species thought to be secure, some unsettling patterns may appear.

References

Bailey, J. W. 1946. The Mammals of Virginia. Williams Printing Company, Richmond, Virginia.

Balcom, B. J., and Yahner, R. H. 1996. Microhabitat and landscape characteristics associated with the threatened Allegheny woodrat. Conservation Biology 10:515–525.

Butchkoski, C. M. 2003. Eastern woodrat research/management. Job code 71801, Annual report to Pennsylvania Game Commission, Harrisburg, Pennsylvania.

Butchkoski, C. M. 2006. Allegheny woodrat research/management. Job Code 71801-05z, Appendix A, Allegheny woodrat habitat site survey code manual. Pennsylvania Game Commission, Harrisburg, Pennsylvania. Available http://www.pgc.state.pa.us/pgc/lib/pgc/reports/2006˙wildlife/71801-05.pdf.

Cudmore, W. W. 1983. The distribution and ecology of the eastern woodrat, *Neotoma floridana*, in Indiana. Ph.D. Dissertation, Indiana State University, Terre Haute, Indiana.

Cudmore, W. W. 1985. The present distribution and status of the eastern woodrat, *Neotoma floridana*, in Indiana. Proceedings of the Indiana Academy of Science 94:621–627.

Federal Register. 1989. 50 CFR Part 17, Endangered and Threatened Wildlife and Plants; Animal Notice of Review. Friday, January 6, 1989. Proposed Rules. Department of the Interior, Fish and Wildlife Service. Federal Register, 54(4):554–579.

Feller, D. J., MacIvor, L., and Bartgis, R. 1994. Population studies of the Allegheny Woodrat (*Neotoma floridana magister)* in Maryland. Report to the Maryland Natural Heritage Program, Maryland Department of Natural Resources, Annapolis, Maryland.

Ford, W. M., Castleberry, S. B., Mengak, M. T., Rodrigue, J. L., Feller, D. J., and Russell, K. R. 2006. Persistence of Allegheny woodrats *Neotoma magister* across the mid-Atlantic Appalachian Highlands landscape, USA. Ecography 29:745–754.

Franz, D., and Slifer, D. 1971. Caves of Maryland. Maryland Geologic Survey Educational Series No. 3, Baltimore, Maryland.
Handley, C. O., Jr., and Patton, C. P. 1947. Wild Mammals of Virginia. Commonwealth of Virginia Commission of Game and Inland Fisheries, Richmond, Virginia.
Hanski, I., and Gilpin, M. 1991. Metapopulation dynamics: brief history and conceptual domain. Biological Journal of the Linnaean Society 42:3–16.
Hassinger, J. D., and Dunn, J. 1990. Job code 06718, Annual report to the Pennsylvania Game Commission, Harrisburg, Pennsylvania.
Hickie, P. F., and Harrison, T. 1930. The Allegheny woodrat in Indiana. American Midland Naturalist 12:169–174.
Hunter, M. L. 1996. Fundamentals of Conservation Biology. Blackwell Science, Malden, Massachusetts.
Johnson, S. A. 2002. Reassessment of the Allegheny woodrat (*Neotoma magister*) in Indiana. Proceedings of the Indiana Academy of Science 111:56–66.
Lay, D. W., and Baker, R. H. 1938. Notes on the home range and ecology of the Attwater wood rat. Journal of Mammalogy 19:418–423.
MacKenzie, D. I., Nichols, J. D., Royle, J. A., Pollock, K. H., Bailey, L. L., and Hines, J. E. 2006. Occupancy Estimation and Modeling: Inferring Patterns and Dynamics of Species Occurrence. Academic Press, Burlington, Massachusetts.
McGowan, E. M. 1993. Experimental release and fate study of the Allegheny woodrat (*Neotoma magister*). Federal Aid Project W-166-E, Job No. VIII-7, Research report to the New York Department of Environmental Conservation, Albany, New York.
Meffee, G. K., and Carroll, C. R. 1997. Principles of Conservation Biology. Second edition. Sinauer Associates, Sunderland, Massachusetts.
Mengak, M. T. 2000. Additional field longevity record for the Allegheny woodrat (*Neotoma magister*). Banisteria 15:50–51.
Mengak, M T. 2002a. Reproduction, juvenile growth and recapture rates of Allegheny woodrats (*Neotoma magister*) in Virginia. American Midland Naturalist 148:155–162.
Mengak, M. T. 2002b. Analysis and summary of eleven years of Allegheny woodrat trapping data in southwest Virginia, 1990–2000. Research report to the Virginia Department of Game and Inland Fisheries, Richmond, Virginia.
Mengak, M. T., Castleberry, S. B., Ford, W. M., Castleberry, N. L., and Rodrigue, J. L. 2002. Longevity for a wild Allegheny woodrat (*Neotoma magister*) in West Virginia. Virginia Journal of Science 53:167–170.
Otis, D. L., Burnham, K. P., White, G. C., and Anderson, D. R. 1978. Statistical inference from capture data on closed animal populations. Wildlife Monographs No. 62.
Paradiso, J. L. 1969. Mammals of Maryland. North American Fauna 66:1–193.
Parkinson, E. A., Berkowitz, J., and Bull, C. J. 1988. Sample size requirements for detecting changes in some fisheries statistics from small trout lakes. North American Journal of Fisheries Management 8:181–190.
Poole, E. L. 1940. A life history sketch of the Allegheny woodrat. Journal of Mammalogy 21: 249–270.
Richards, R. L. 1987. The Quaternary distribution of the eastern woodrat, *Neotoma floridana*, in Indiana. Proceedings of the Indiana Academy of Science 96:513–521.
Thompson, W. L. (ed.). 2004. Sampling Rare or Elusive Species. Island Press, Washington, D.C.
Thompson W. L., White, G. C., and Gowan, C. 1998. Monitoring Vertebrate Populations. Academic Press, Burlington, Massachusetts.
White, G. C., Anderson, D. R., Burnham, K. P., and Otis, D. L. 1982. Capture-recapture and removal methods for sampling closed populations. Los Alamos National Laboratory Publications LA-8787-NERP, Los Alamos, New Mexico.

Chapter 8
Managing Surface Rock Communities for *Neotoma magister*

Jerry D. Hassinger, Calvin M. Butchkoski, and Duane R. Diefenbach

Introduction

The Allegheny woodrat (*Neotoma magister*) is, perforce, limited to rocky habitats (Castleberry et al. 2006) located in forest interiors. This association with surface rock islands in large tracts of contiguous forest means suitable habitat always has been rare and is becoming more so. Today, *N. magister* is endangered or threatened in six of the 12 states that encompass its range (Wright Chapter 1). Numerous authors, most recently Ford et al. (2006) and LoGiudice (2006, Chapter 2), suggest multiple stressors have contributed to a protracted population decline and local extinctions of *N. magister* populations. We consider these stressors as the result of two basic threats.

First, forest fragmentation likely has provided more habitat generalists with access to rock outcroppings, and resulted in more predators and increased competition for food and cover. Raccoons (*Procyon lotor*) pose an additional threat by hosting a roundworm parasite (*Baylisascaris procyonis*) lethal to *N. magister* (McGowan 1993, LoGiudice 2000). Forest fragmentation also likely has compromised dispersal, a condition necessary for the persistence of insular *N. magister* subpopulations and metapopulations.

Second is the degradation or elimination of traditional foods, primarily storable mast, as a result of the loss of American chestnut (*Castanea dentata*), erratic acorn production following oak (*Quercus* spp.) defoliation by gypsy moths (*Lymantria dispar*), and changes in tree species composition that reduce oak abundance (Diamond et al. 2000, Rodewald 2003). Silvicultural practices, a thriving white-tailed deer population (*Odocoileus virginianus*), acid deposition, and suppression of wildfires also likely have adversely affected the amount and quality of foods available to *N. magister.*

These two major threat categories are common to all regions encompassing the distribution of *N. magister*, and are fundamental to the recommendations we provide

J.D. Hsaainger
Pennsylvania Biological Survey, 105 Adams Lane, Millersburg, PA 17061

J.D. Peles, J. Wright (eds.), *The Allegheny Woodrat*, DOI: 10.1007/978-0-387-36051-5_8,

in this chapter. Some individual stressors (e.g., oak decline, droughts, and human development on ridges) are an increasing problem throughout the range of *N. magister*. Other stressors are important in the northern third of the range of *N. magister*, such as energetic demands associated with cold weather, gypsy moth defoliation and loss of oak trees, and raccoon roundworm.

In this chapter, we provide recommendations for the management of the Allegheny woodrat. Threats to the welfare of *N. magister*, notably forest fragmentation and changing tree species composition, are ominous and continue almost unabated in their degradation of habitat for this species. Consequently, our management recommendations are designed to avoid, mitigate, or compensate for these primary threats and their related stressors as quickly as possible. However, we acknowledge the need to improve our understanding of the factors threatening this species, and we strongly encourage that implementation of any of our recommendations be accompanied by appropriate monitoring and research. Also, we urge greater federal and state interagency cooperation and consistency in defining and delineating habitat sites, subpopulations, and metapopulations.

Our approach to identifying components of a management plan for the Allegheny woodrat was to focus on the protection of surface rock communities, because the misleading and pejorative name "woodrat" immediately compromises public and political support for single species management. Numerous charismatic vertebrate species are associated with these ecosystems, including over-wintering black bear (*Ursus americanus*), bobcat (*Felis rufus*), raven (*Corvus corax*), and peregrine falcon (*Falco peregrinus*). We have identified at least 27 species associated with surface rock habitats that are endangered, threatened, or of conservation concern in Pennsylvania (PGC-PFBC 2005). Management of biological communities associated with rock outcrops in a multi-species context is more efficient, easier to justify, and potentially entails less derision, than the management of a single species. Therefore, the recommendations in this chapter complement, to the extent possible, a community approach to the management of surface rock habitats. In addition, we encourage the scientific community and resource managers to find a more socially acceptable common name for the Allegheny woodrat. Therefore, we hereafter refer to this species as *N. magister*.

Definitions for *Neotoma* Populations and Habitat

For monitoring *N. magister* in Pennsylvania, we have developed a habitat and population classification system (Butchkoski 2006) based on its social organization, habitat use, dispersal ecology, and metapopulation dynamics.

Surface Rock Communities

We refer to cliffs, rock outcrops, talus slopes and caves, and the forested habitat immediately surrounding these areas, as *surface rock communities* because of their unique physical features that provide ledges, rock overhangs, and multiple crevices leading to deep, frost-free, underground galleries. Such features are critical to the presence and persistence of many species, including *N. magister*, associated with this community type. The forests surrounding surface rock communities are an equally important habitat component because plants growing in or around surface rock provide the food sources for *N. magister*.

Activity Center

At the smallest scale, we use the term *activity center* to represent the site of a breeding female's den area. Operationally, we define an activity center to be an area, approximately 30 m in diameter, that is centered on signs of fresh or old activity in surface rock (toilet areas and food caches) or, if no sign is present, an area containing suitable surface rock for a potential den site. The central coordinates of an activity center are recorded along with all sign within a 15 m radius to facilitate the monitoring of occupancy. Activity centers are classified as active (evidence of current occupation), inactive (evidence of former occupation), historical (*N. magister* were known to be present in the area but no evidence of current or former occupancy), and potential (contains rock crevices that likely provide access to frost-free belowground cavities but no evidence of occupation or historical records).

Habitat Site

On a slightly larger scale, we use the term *habitat site* to approximate the patch of habitat that can support an *N. magister* subpopulation within a metapopulation. Operationally, a habitat site is a surface rock island, or a cluster of islands, containing one or more activity centers and separated from adjacent habitat sites by $\geq$200 m of unsuitable or nonrock habitat. The area of a habitat site is defined by one or more rectangles that encompass the surface rock plus a surrounding 200-m wide buffer to represent potential foraging area. Each active habitat site houses an *N. magister* subpopulation which may contain one or more breeding assemblages (a female, offspring, and a visiting male). We calculate a crude estimate of population size of a habitat site as the number of activity centers with fresh sign multiplied by five. We consider habitat sites to be the basic management units for *N. magister* (Fig. 8.1).

Metapopulation Area

We use the term *metapopulation area* to indicate our best approximation of the area that encompasses a set of spatially discrete *N. magister* subpopulations connected via dispersal corridors. Maximum dispersal distances reported for *N. magister* are 6 km (Wright 1998) and 2.5 km (Castleberry et al. 2001, Mengak 2002). Using 10 km as an approximate maximum distance for dispersal, we include in a given

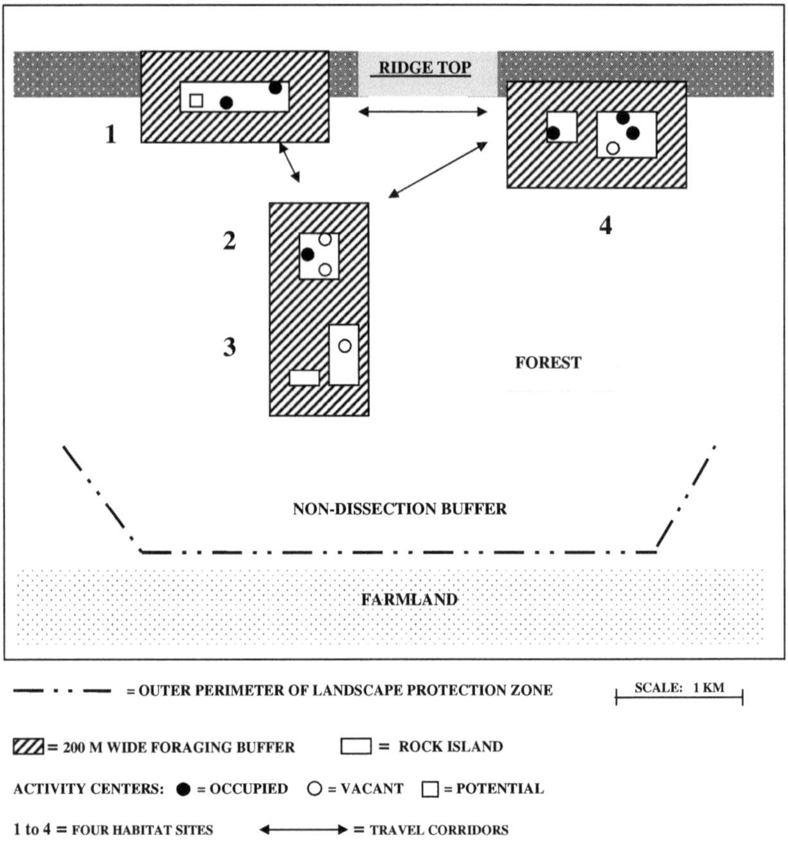

Fig. 8.1 Hypothetical illustration of four *Neotoma magister* habitat sites to demonstrate terminology recommended in this chapter. Habitat site #3, closest to farmland, was used in the past, but is now vacant. Rock islands in habitat site #4 are not separated by ≥200 m of primarily nonrock habitat; hence they are included in the same habitat site

metapopulation area only those habitat sites less than 10 km from another site, but we also consider terrain features to identify probable impermeable dispersal barriers. We recognize our distance criterion may include some habitat sites within a single metapopulation area that are effectively isolated. Nonetheless, we believe this definition is relevant to developing management plans and protecting habitat.

Management Issues and Recommendations

State-specific economic, social, and political constraints are likely to influence the implementation of landscape-level management recommendations for *N. magister*. However, state-specific recommendations for protection of habitat sites and metapopulations will provide benefits if developed within the context of existing regional plans and recovery planning guidance for endangered species (National Marine Fisheries Service 2006). For example, the Central Appalachian Forest Ecoregional Plan (Nature Conservancy 2001) should be consulted when developing a management plan in Maryland, Pennsylvania, Virginia, and West Virginia. Other regional plans exist for North American landbirds (Rich et al. 2004) and southern Appalachian forests (Irwin et al. 2002). Moreover, we recommend the use of adaptive resource management (Salafsky et al. 2001), where data collected during monitoring is used to assess the outcomes of management actions. An adaptive management approach is important because of the regional uncertainty regarding the relative importance of specific stressors to the decline of *N. magister*, and because the outcomes of management actions may not be evident for decades (Nichols et al. 1995). Also, pervasive issues like global climate change, air pollution, energy independence, and invasive species (inclusive of pathogens and pests) will likely further contribute to decision uncertainty.

We identify and give background information on five issues important to the viability of *N. magister*. We provide recommendations that, if implemented in concert with other actions, are reasonably expected to lead to the eventual restoration, maintenance, or delisting of *N. magister*. Conservation and management actions should focus on avoiding, reversing where possible, or otherwise alleviating the most important threats that compromise the persistence of *N. magister*. Consequently, on a state-by-state basis, it will be essential to identify the most critical stressors contributing to this species' decline, and identify the combination of habitat sites and metapopulations where implementation of conservation measures is reasonably expected to lead to a timely and most cost-effective restoration, maintenance, or delisting of *N. magister*. Conversely, it will be important to identify ecological traps and sink habitat.

Monitoring Habitat and Populations

State-specific surveys and delineation of habitat sites critical to the welfare of the *N. magister* are far from complete and in some instances nonexistent. Northern states with documented population declines (Indiana, Maryland, New Jersey, New York, Ohio, Pennsylvania, and Virginia) have some of the best information, but the definition and delineation of sites among states has been inconsistent. In this chapter, we have provided definitions for habitat site and metapopulation area, and recommend their general use. Protection and restoration of *N. magister* populations require knowledge of extant and historical habitats, the present condition of these sites, and their relationships in a metapopulation context (Runge et al. 2006).

Habitat and population monitoring can provide important information on population trends, occupancy of activity centers or habitat sites, and estimates of rates of extinction and colonization (MacKenzie et al. 2006). Knowledge of habitat sites can aid in the identification of conservation priorities (e.g., whether active habitat sites are located on public or private land). Habitat- and species-oriented surveys are often time consuming and expensive, but survey techniques can be developed that do not require live-trapping (e.g., occupancy models). State and federal agencies and land stewards will require this information to develop and prioritize management actions.

Our experience in Pennsylvania provides some insight regarding the benefit of monitoring *N. magister* populations and conducting habitat-oriented surveys. From 1982 to 2006, employees of the Pennsylvania Game Commission surveyed 802 habitat sites; 197 habitat sites contained woodrat sign but were inactive, and 246 habitat sites contained at least one active activity center (Mengak et al. Chapter 7). The total area of occupancy of the 246 active habitat sites was 40 km^2, and 87% of all habitat sites were $\leq$5 ha. Risk of extinction at these sites is high because they are small and distributed among 31 metapopulation areas (Belovsky et al. 1994).

The World Conservation Union Red List Criteria (IUCN 2001) classify a taxon as endangered when: (1) the area of occupancy is less than 500 km^2; (2) the population is composed of disjunct subpopulations; and (3) there is a continuing decline in the area of occupancy or quality of habitat. The status of *N. magister* in the northern third of its range appears to satisfy all the three criteria. If status determination and monitoring efforts in southern states produce similar results, *N. magister* should be considered for endangered species status at the federal level, which would enhance legal protection and financial support for this species.

Recommendations

(1) Adopt standardized definitions for activity centers, habitat sites, and metapopulation areas throughout the range of *N. magister*. We recommend the definitions provided in this chapter and detailed in the Pennsylvania Game Commission Allegheny Woodrat Habitat Site Survey Code Manual (Butchkoski 2006).

(2) Train field technicians to accurately identify sign of *Neotoma* activity, consistently delineate and map habitat sites and activity centers, and accurately classify activity centers as active or inactive. Regional workshops would ensure consistency among state agencies.

(3) Implement assessment methods that consider the limited resources and personnel available for population surveys such as occupancy estimation and modeling of activity centers (MacKenzie et al. 2006). Using occupancy modeling, it may be possible to complete statewide surveys within five years and incorporate monitoring for other species associated with these habitats (e.g., timber rattlesnake, *Crotalus horridus*).

(4) Prioritize habitat sites and metapopulation areas for protection and management actions (see Appendix), and transfer these results to administrators and land stewards.

Forest Fragmentation

Existing habitat sites, including short (<2 km) dispersal corridors, require protection from human-caused, permanent forms of forest fragmentation and dissection

(Hunter 1996). Along with habitat loss, habitat fragmentation is one of the greatest threats to native wildlife worldwide (Wilcox and Murphy 1985, Noss 1987, Wilcove 1987, Noss and Cooperrider 1994). Dissection is a form of fragmentation, but we believe it deserves emphasis in the context of *N. magister* habitat because roads, utility right-of-ways, wind energy generation plants, and communication towers are contributing to the dissection and eventual fragmentation of forests surrounding *N. magister* habitats. Stressors affecting *N. magister* (disease, predation, and competition for food resources) are likely exacerbated by forest fragmentation and dissection because Balcom and Yahner (1996) reported greater percentages of cropland and other agricultural land use at unoccupied historical sites than at currently occupied sites.

Using data from Pennsylvania's statewide monitoring of *N. magister* habitats, we identified 439 habitat sites (active, inactive but containing old sign, or known to historically contain *N. magister*) in 79 metapopulations. We classified each site as occupied or not, measured the distance to the nearest nonforest land cover (≤1 km, 1–2 km, and >2 km) greater than 100 ha, and measured the size of the habitat site. We used logistic regression to model the probability that a habitat site was occupied as a function of distance from nonforest habitat and size of the habitat site. We used odds ratios (with 95% confidence intervals) to express the relative likelihood that a habitat site was occupied. We found habitat sites were more likely to be occupied the farther they were located from nonforest land cover (Fig. 8.2), and larger habitat sites were more likely to be occupied. Habitat sites located >2 km from nonforest edge were 7.1 times more likely to be occupied than sites ≤1 km from nonforest edge (95% CI = 4.1–12.3), and sites >1–2 km from a nonforest edge were 3.9 times more likely to be occupied (95% CI = 2.1–7.2). For every 5 ha increase in area, a habitat site was 1.2 times more likely to be occupied (95% CI = 0.97–1.49). These data are simply snapshots at one point in time, and monitoring programs would be necessary to estimate extinction and colonization rates (MacKenzie et al. 2006).

Forest fragmentation has been increasing since the early 1900s as more landowners own smaller parcels of forestland (Mehmood and Zhang 2001), a situation that makes it increasingly difficult to sustainably manage forest-related resources (Crim et al. 2002). Confounding any sense of urgency, however, is the delay between habitat changes and resulting species loss (Tilman et al. 1994). As habitat

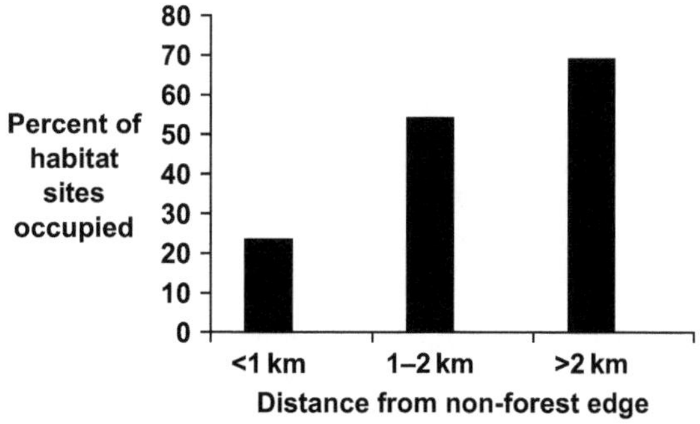

Fig. 8.2 Percent of habitat sites occupied by *N. magister* as a function of distance from the nearest nonforest habitat for 439 habitat sites in Pennsylvania

becomes fragmented, a suite of interacting stressors evolves, and years or decades later a metapopulation goes extinct. Although habitat loss and fragmentation are immediate and obvious, their effect on wildlife populations may be difficult to measure or detect until years later. LoGiudice (2006, Chapter 2) explored the major hypotheses for the decline of *N. magister*, and what emerged was a complicated history involving multiple, relatively minor stressors that were largely attributable to fragmentation and dissection of forests.

Landscape-level forest fragmentation is the result of encroaching farmlands, surface mining, human developments, and transportation and utility corridors. The effects of large-scale, permanent openings next to or within forest communities are exacerbated by additional dissections such as secondary roads, unimproved roads, communication towers, and windmill sites (Fig. 8.3). In addition to the perennial disturbance associated with fragmentation and dissection of forests, there is the transitory disturbance resulting from forest management (e.g., tree harvest, access roads, etc.).

Goodrich et al. (2006) reported that Pennsylvania is more than 60% forested but 58% of the forest is within 100 m of a road or nonforest habitat. Forest fragmentation in Pennsylvania continues, especially on ridge tops, because of the proliferation of communication towers, sites for generation of wind energy, and the right-of-ways required to service these sites. Similarly, if recent (2001–2004) forest loss trends continue, Virginia will lose more than one million acres of forest land in the next 25 years (Virginia Department of Forestry 2006). It is likely that changes

Fig. 8.3 (**A**) *Neotoma magister* habitat site in the Ridge and Valley Province of Pennsylvania. Examples of impacts on the site include (**B**) fragmentation by fields and human development, and (**C**) dissection by road and utility right-of-way

observed in Pennsylvania and Virginia are typical of most states throughout the range of *N. magister.*

Fragmentation increases the abundance of habitat generalists and allows them to penetrate remnant interior forests (Fig. 8.3). White-tailed deer, black bear, and wild turkey (*Meleagris gallopavo*), for example, thrive in fragmented habitats, but also use forests where they compete with *N. magister* for acorns (McShea and Schwede 1993, McShea 2000, Feldhamer 2002). White-tailed deer, rare in the late 1800s and abundant for much of last century, have degraded and simplified the forest understory (Latham et al. 2005). Myers (1997) suggested that combined with aggressive fern (*Dennstaedtia punctilobula* and *Thelypteris noveboracensis*) growth, deer may reduce the availability and nutritional quality of the ground cover vegetation essential for *N. magister* winter food caches.

Raccoon densities in Pennsylvania are as much as three times greater (18.0–41.4 per km^2) in fragmented habitats compared to forested areas (5.1–11.9 per km^2; Hayden 1988). Surface rock communities closer to farmlands likely receive more use by raccoons, whose feces may contaminate habitat sites with *B. procyonis*, which is fatal when ingested by *N. magister* (LoGiudice 2000). The contamination of habitat sites with this parasite can create population sinks (McGowan 1993). We believe conserving or restoring contiguous forest will reduce raccoon densities near *N. magister* habitat sites and potentially reduce the chances of site contamination by raccoon roundworm. On occupied habitat sites, where the raccoon roundworm is a known mortality factor, the use of antihelmintics shows promise (LoGiudice 1995), but more research is needed. Predators of *N. magister* thrive in fragmented habitats (Soulé 1990, Garrott et al. 1993, Oehler and Litvaitis 1996). Black rat snake (*Elaphe obsoleta*), great horned owl (*Bubo virginianus*), feral cat (*Felis catus*), raccoon, skunk (*Mephitis mephitis* and *Spilogale putorius*), fox (*Vulpes vulpes* and *Urocyon cinereoargenteus*), and long-tailed weasel (*Mustela frenata*) are species documented as preying on *N. magister* and *N. floridana* (Poole 1940, Fitch and Rainey 1956, Wink et al. 1987).

Managing the competitors, parasites, and predators that result in increased mortality rates, lower reproductive rates, and population isolation is not biologically or financially feasible. However, we believe minimizing or reversing forest fragmentation is a more efficient and cost-effective approach because most stressors are associated with fragmented habitats. Similarly, Schneider (2001) concluded predator management is not a viable strategy to combat the threat to survival of endangered prey, but that careful planning of the landscape pattern could compensate for predation effects.

Recommendations

(1) Manage public lands to protect forest interior habitat and surface rock communities. Acquire land easements to protect surface rock habitat on privately owned land, especially because this habitat type usually is not commercially valuable. However, it is essential that the forested habitat within 2 km of a

habitat site not be fragmented. This caveat may require a 2-km buffer on one or more sides of the habitat site if there is a potential for development.

(2) Use a multiple use module (MUM; Noss and Harris 1986) to guide allowable uses and prioritize management activities. A model *N. magister* management MUM consists of an undisturbed habitat site as the core, surrounded first by a 200-m-wide foraging zone where management actions are designed to improve production of hard mast, and then by an outer 2-km-wide landscape protection zone or a nondissection buffer where commercial forest management may occur, but roads, right-of-ways, and other forms of permanent fragmentation and dissection are discouraged (Fig. 8.1).

(3) Prevent the permanent dissection of the intervening forest when active habitat sites are within 2 km of a large area (≥100 ha) of nonforest (Figs. 8.1 and 8.3).

(4) Maintain forested and dissection-free dispersal corridors between habitat sites <2 km apart. Short dispersal distances are more common, and most successful interhabitat site dispersal rarely exceeds 2.5 km.

(5) Retire and reforest timber sale haul roads within 1 km of occupied habitat sites and main haul roads that intersect or parallel dispersal corridors ≤2 km in length. Maintain as much of a closed tree canopy as possible for the above all existing dirt roads within 2 km of occupied habitat sites to minimize habitat changes that benefit generalist species.

(6) Observe fragmentation guidelines to reduce the probability of habitat sites becoming contaminated with raccoon feces containing raccoon roundworm. On occupied habitat sites where the roundworm is a known mortality factor, initiate research to determine why *N. magister* still occupies the site, and test the efficacy and feasibility of using antihelmintics as part of a program to eliminate the parasite.

(7) Use bat-friendly gates to protect cave and mine entrances known to be used by *N. magister*. Bat Conservation International (Austin, Texas, http://www.batcon.org) has information on gate designs and provides workshops on gating techniques. Avoid creating permanent access (roads, utility right-of-ways, and trails) to the vicinity of caves and mines, known rattlesnake basking sites, or currently active *N. magister* habitat sites.

Food Availability and Quality

N. magister is primarily an opportunistic herbivore with a catholic diet (Castleberry et al. 2002, Castleberry and Castleberry Chapter 6). This species eats and stores, presumably for later consumption, plants, leaves, seeds, fruit, and fungi. The quantity and quality of food available to *N. magister* has deteriorated over the past century because of changes in forest conditions. Unfortunately, most studies of *N. magister* food habits have been conducted on populations experiencing a long-term decline. Changes since 1880 in eastern North America forests have resulted in a sustained reduction in the quantity and quality of food available to *N. magister* (Diamond et al. 2000, McShea and Healy 2002).

Despite *N. magister's* catholic diet, hard mast crops are likely important food sources because of their energy content and potential for long-term storage. As discussed in Chapter 2, extensive timber extraction during 1880–1920 caused major changes in the age structure and species composition of Appalachian forests, and the chestnut blight (*Cryphonectria parasitica*) resulted in the loss of the American chestnut. The replacement of American chestnut by oaks resulted in less consistent mast production, which has been further exacerbated by defoliation and tree mortality caused by the gypsy moth caterpillar. Today, forest tree species composition is changing from oaks to species that do not produce mast crops (e.g., black birch, *Betula lenta* and red maple, *Acer rubrum*). In addition, competition with other wildlife (e.g., white-tailed deer, black bear, and wild turkey) for hard mast and changes in understory composition caused by deer browsing, fire suppression, invasive exotic plant species, and competitive exclusion by plants tolerant of acid soils (e.g., ferns and red maple) have further reduced the diversity and abundance of food resources important to *N. magister*.

The abundance and distance of a preferred food item from a den site dictate the time and energy required to forage. Predation risk increases when individuals travel farther searching for and gathering food and net energy gained is lower. The two northernmost *Neotoma* species, *N. cinerea* in the Canadian Rockies and *N. magister* in the Appalachians, are dependent on islands of surface rock for den sites and have larger home ranges than other *Neotoma* species (Castleberry Chapter 4). Castleberry et al. (2001) and Topping and Millar (1996) speculated that food resources within short foraging distances of rock outcrops may be insufficient, requiring these species to move greater distances (as much as 500 m for *N. magister*) than other species of *Neotoma*. But when forests are fragmented, it may no longer be safe for *N. magister* to regularly move great distances for food.

The positive relationship between mast production and reproductive output has been noted for many species of rodents (Gorman and Roth 1989, Ostfeld et al. 1996, McShea 2000), and *N. magister* populations should respond in a similar manner with increased reproduction following years of abundant mast production, especially acorns. Because acorn production varies annually, managers should strive to provide a diversity of food categories and food species within habitat sites. Food categories described by Castleberry et al. (2002) include fungi and lichens, ferns, buds and leaves, other green vegetation (primarily forbs), hard mast, soft mast, other seeds, and insects. Rubino and McCarthy (2003) identified 100 species of macrofungi on oak logs in Ohio and species richness increased with log volume. A diversity of tree species of appropriate age is equally important. For example, *Russula* spp., a macrofungus found in *N. magister* middens (Poole 1940, Richmond and Rosland 1949) is an ectomycorrhizal associate of pine (*Pinus* spp.) and birch (*Betula* spp.) trees.

Numerous studies of wild rodents have shown that supplementary feeding results in a positive population response (Gilbert and Krebs 1981, Dobson and Oli 2001) by increasing reproduction and survival. Supplemental feeding is intensive, expensive, and often a procedure of last resort, although it provides an opportunity to engage volunteers. Supplemental feeding should be conducted only in conjunction with research in an adaptive management framework to test hypotheses regarding limiting factors, and to ensure that subsequent supplemental feeding results in intended consequences.

Recommendations

(1) Manage trees to increase the production of hard mast within the habitat site (Hornsby et al. 2005; Fig. 8.1). Thinning other species (e.g., red maple) may release oaks, but increasing acorn production is a challenging task. We recommend Johnson (1994) and McShea and Healy (2002) for details.
(2) Leave woody material, including felled trees, within a habitat site. Scattered islands of down wood, including many decay classes and sizes, ranging from twigs to logs, provide protection and fungal habitat.
(3) Beyond the foraging buffer and within the landscape protection zone (Fig. 8.1), use long rotations and variable-age tree retention harvesting techniques (Curtis 1997, Franklin et al. 1997, Tappeiner et al. 1997) to maximize the production of hard mast. At least two-thirds of the foraging buffer's outer perimeter should abut a stand of mature mast-producing trees at any point in time.
(4) Plant native tree species and produce mast primarily within 200 m of activity centers. Rather than seedlings, we recommend planting container stock because of the expense of accessing habitat sites and the poor growing conditions (shallow soils with poor acid buffering capacity and xeric conditions). Protecting plantings from deer browsing may be necessary. We recommend Fowells (1965) and Burns and Honkala (1990) for propagation and site requirements for the growth of suitable tree species.
(5) Cooperate with American Chestnut Cooperators Foundation (http://www.accf-online.org/index.html) and the American Chestnut Foundation (http://www.acf.org/) to restore the American chestnut to eastern hardwood forests. The American chestnut is a reliable mast producer, unlike the sporadic production that characterizes the larger oak species.
(6) Strive for a diversity of trees, shrubs, vines, and forbs and fungi within the habitat site. For instance, grape vines (*Vitis* spp.) over rocks provide food, cover, and water. Other choices include native and non-native (if noninvasive) species: American hazel (*Corylus americana*), American mountain ash (*Sorbus americana*), blackgum (*Nyssa sylvatica*), brambles (*Rubus* spp.), Kentucky coffeetree (*Gymnocladus dioica*), mulberry (*Morus* spp.), sassafras (*Sassafras albidum*), scarlet elder (*Sambucus pubens*), Turkish tree hazel (*Corylus colurna*), and winged sumac (*Rhus copallina*). Degraaf and Witman (2002) described the range, habitat, and propagation characteristics for these and 148 other species of trees, shrubs, and vines. For regional or local advice, partner with the North American Fruit Explorers, a network of individuals devoted to the discovery, cultivation, and appreciation of superior varieties of fruits and nuts (http://www.nafex.org).
(7) If necessary and feasible, use supplementary feeding to augment natural foods during critical periods to help a small subpopulation persist or to help establish

a population at a new habitat site. Provide storable foods (either natural or man-made), dispersed throughout rock crevices, between mid-July and late October.

Dispersal and Translocation

Stochastic processes, such as demographic, environmental, catastrophic, and genetic perturbations (Shaffer 1981), result in some unoccupied habitats at any given point in time. Historically, it is likely a vacant site would eventually be reoccupied by dispersing *N. magister.* Long-distance dispersal (≥2 km) can be compromised by: (1) too few dispersers; (2) man-made and natural barriers; (3) predators; (4) large distances between suitable habitat; and (5) emigrating to an ecological trap. However, in the northern third of the range of *N. magister*, current evidence suggests successful long-distance dispersal beyond a habitat site or cluster of adjacent habitat sites has been reduced because more known habitat sites are now vacant. For example, 48 of 79 (61%) metapopulation areas in Pennsylvania are now vacant, and distance (≥10 km) from the nearest source population virtually precludes their recolonization by natural dispersal.

Failed dispersal attempts can have dire consequences. Remnant subpopulations become spatially isolated and vulnerable to demographic and genetic problems that reduce persistence (Belovsky et al. 1994) because isolated vacant sites (Appendix) have no source of immigrants. Thus, translocations of individuals to supplement isolated populations or to re-establish populations in vacant habitat sites may become necessary. Given the status of *N. magister* populations in the northern portion of the species' range, and the fragmented nature of surface rock habitats, repeated or multiple translocations may be necessary to restore extirpated populations or enhance genetic diversity of small populations (Lubow 1996).

Preliminary results of an experimental translocation (Wright 1998) in Pennsylvania and a reintroduction of *N. magister* to a vacant site in New York (McGowan 1993) indicate translocation is feasible but may not be successful. Hence, translocation programs should use an adaptive management approach to evaluate the genetic implications of translocations and incorporate a comprehensive monitoring program. Little is known of the population dynamics of *N. magister*; thus, translocations should be designed to investigate optimal translocation strategies (Lubow 1996) to maximize success.

Recommendations

(1) Map habitat sites and metapopulation areas to help identify potential dispersal corridors. This information could be used to prioritize habitat protection and translocation strategies.

(2) Establish specific goals and objectives and employ a multidisciplinary approach for all stages (planning, preparation, release, and postrelease monitoring) of any translocation or restoration program. The Species Survival Commission of

the World Conservation Union (http://www.iucn.org/themes/ssc/publications/policy/reinte.htm) provides guidelines for reintroductions, and Serfass (Chapter 10) provides specific recommendations for *N. magister*.

Surface Rock Community Management

Current community classification systems are inadequate for describing surface rock communities. For example, "barrens" is a misleading term for a surface rock community with its associated vegetation. Surface rock communities are at risk because generally they have not been recognized and delineated. Surface rock is a term of convenience used to describe insular habitat that is readily identified by resource managers and private landowners. We refer to exposed rock in the form of cliffs and ledges, outcrops, rock slides and talus, rocks and boulders, boulder fields, float blocks, cave and mine entrance zones, and old rock quarries as surface rock. Surface rock is multidimensional having length, width, and depth. The depth of rock habitat is important because the crevices, galleries, and caves that provide access below the frost line are critical for some species (e.g., rattlesnakes and *N. magister*).

Surface rock communities also include the ecotone with the surrounding forest vegetation. Managing the surface rock community and an associated suite of focal species is more efficient, easier to justify, and will engender greater public support than the management of *N. magister* per se. Furthermore, potentially conflicting recommendations to improve habitat for different species can be integrated and prioritized.

In Pennsylvania, we identified at least 27 endangered, threatened, and species of conservation concern (PGC-PFBC 2005) associated with surface rock communities. Stonecrops (*Sedum* spp.), rock tripe (*Umbilicaria mammulata*), calciphilic snails (*Gastrocopta* spp.), green salamanders (*Aneides aeneus*) rattlesnakes, rock voles (*Microtus chrotorrhinus*), and *N. magister* are dependent upon surface rock ecosystems for one or more of their life requirements. Also, species that are not wholly dependent on surface rocks include over-wintering black bears, nesting vultures, and 27 species of amphibians and reptiles that use terrestrial habitats (J. Hassinger unpublished data). The recommendations in this chapter for protecting habitat for *N. magister* will benefit forest interior species such as the cerulean warbler (*Dendrocia cerulea*) and timber rattlesnake, and may help reduce disturbance and the spread of invasive species. Furthermore, given the limited extent of occupied-and-buffered (to 2 km) *N. magister* habitat sites, the recommendations in this chapter, if implemented, should have no adverse effect on populations of common and widespread species of wildlife.

Recommendations

(1) Survey, classify, map, and monitor surface rock communities, including those that are not known to support existing *N. magister* populations. The absence

of *N. magister* does not preclude the existence of other focal species (e.g., small-footed myotis, *Myotis leibii*).

(2) Minimize any form of forest fragmentation on public forests, to the extent possible, because limiting or reducing fragmentation on private lands will be more difficult and expensive. This will benefit multiple special concern species, particularly forest interior birds.

(3) Coordinate management plans for surface rock communities and focal species (e.g., *N. magister*) with larger-scale plans, such as the North American landbirds (Rich et al. 2004) and the southern Appalachian forests (Irwin et al. 2002). These plans acknowledge that "On public lands, land use plans must be based on sustaining or restoring long-term biotic integrity of ecosystems" (Rich et al. 2004).

Conservation and Management of Declining Species: Shared Stewardship

The responsibility for the conservation and management of *N. magister* and surface rock communities is shared by 12 states and the federal government. Prescriptions for stabilizing and recovering *N. magister* metapopulations are straightforward: (1) prevent and reverse, where possible, fragmentation and dissection of forests within metapopulation areas; (2) improve the quantity and quality of food near den sites; and (3) consider direct feeding and assisted dispersal for qualifying habitat sites (Appendix). These fundamental prescriptions, however, are difficult to implement because population surveys are incomplete, forest fragmentation is likely to increase, and establishing self-sustaining food resources at habitat sites could take decades. We provide specific recommendations for addressing management issues and encourage resource managers to incorporate these recommendations into state Comprehensive Wildlife Conservation Plans and existing National Park and Forest management plans.

In 2005, every state developed a Comprehensive Wildlife Conservation Plan, also called a Wildlife Action Plan (http://www.teaming.com). These plans are required to receive federal funds to proactively conserve species of concern and vital natural areas, such as surface rock communities, before they become rare and require extraordinary measures and expense for their protection. These plans are habitat-oriented, and as such, recommendations to improve habitat for *N. magister* will (1) benefit other special concern species; (2) reduce the spread of invasive species; and (3) protect special habitat for charismatic species (e.g., cliff-nesting peregrine falcons and over-wintering black bears).

Collectively, Wildlife Action Plans list varying levels of concern for *N. magister*, which could benefit from interstate cooperation. For example, New York has extensive tracts of public land and numerous vacant *N. magister* habitat sites. Similarly, Pennsylvania and Maryland have a mix of occupied and unoccupied sites, many of which are on public land. Biologists in these states, where *N. magister* is in decline, have an opportunity to study factors that contribute to source/sink habitats and experiment with translocations and restoration strategies within an adaptive management framework. On the other hand, apparently stable populations of *N. magister* reside in West Virginia, Virginia, Tennessee and Kentucky, and resource managers in these states share a responsibility for the welfare of this species and surface rock communities.

Despite the loss and continuing degradation of contiguous forest, we do not advocate ignoring the management of privately owned woodlands. The management of public forests, however, is likely the key to the future existence of *N. magister* and other forest interior species. Public lands are important because they contain some of the largest remaining tracts of contiguous forest within the Central and Southern Appalachian Mountains. Consequently, management of public forests must compensate for expected continual fragmentation and loss of interior forest on private lands.

The insular and patchily distributed nature of surface rock habitats, and regionally specific threats, dictate research and management priorities. Addressing these priorities in a comprehensive and coordinated manner will require the cooperation of research institutions, state and federal agencies with land management responsibilities and regulatory authority, and nongovernmental organizations. We recommend the formation of a representative oversight team to initiate such cooperation. Paraphrasing the goal of Partners in Flight (http://www.partnersinflight.org), a useful model, the missions of this team would be to: (1) coordinate and focus resources through public and private sector efforts; and (2) improve monitoring and inventory, research, management, and educational programs that affect special concern species associated with surface rock communities throughout the range of *N. magister*.

References

Balcom, B. J., and Yahner, R. H. 1996. Microhabitat and landscape characteristics associated with the threatened Allegheny woodrat. Conservation Biology 10:515–525.

Belovsky, G. E., Bissonette, J. A., Dueser, R. D., Edwards, Jr., T. C., Luecke, C. M., Ritchie, M. E., Slade, J. B., and Wagner, F. H. 1994. Management of small populations: concepts affecting the recovery of endangered species. Wildlife Society Bulletin 22:307–316.

Burns, R. M., and Honkala, B. H. 1990. Silvics of North America: 1. Conifers. 2. Hardwoods. Agriculture Handbook 654, USDA Forest Service, Washington, D.C.

Butchkoski, C. M. 2006. Allegheny woodrat research/management. Job Code 71801-05z, Appendix A, Allegheny woodrat habitat site survey code manual. Pennsylvania Game Commission, Harrisburg, Pennsylvania. Available http://www.pgc.state.pa.us/pgc/lib/pgc/reports/2006˙wildlife/71801-05.pdf.

Castleberry, S. B., Ford, W. M., Wood, P. B., Castleberry, N. L., and Mengak, M. T. 2001. Movements of Allegheny woodrats in relation to timber harvesting. Journal of Wildlife Management 65:148–156.

Castleberry, N. L., Castleberry, S. B., Ford, W. M., Wood, P. B., and Mengak, M. T. 2002. Allegheny woodrat (*Neotoma magister*) food habits in the central Appalachians. American Midland Naturalist 147:80–92.

Castleberry, S. B., Mengak, M. T., and Ford, W. M. 2006. *Neotoma magister*. Mammalian Species 789:1–5

Crim, S., Dubois, M. R., Finley, J., Fletcher, R., Robinson, L. J., and Blanche, C. 2002. Forest fragmentation extension programming: a national initiative. Available http://sfp.cas.psu.edu/fragmentation/Forest%20Fragmentation.pdf.
Curtis, R. O. 1997. The role of extended rotations. *In* Creating a Forestry for the Twenty-First Century: The Science of Ecosystem Management, eds. K. A. Kohm and J. F. Franklin, pp. 165–170. Island Press, Washington, D.C.
Degraaf, R. M., and Witman, G. M. 2002. Trees, Shrubs, and Vines for Attracting Birds. University Press of New England, Lebanon, New Hampshire.
Diamond, S. J., Giles, R. H., Jr., Kirkpatrick, R. L., and Griffin, G. J. 2000. Hard mast production before and after the chestnut blight. Southern Journal of Applied Forestry 24: 196–201.
Dobson, F. S., and Oli, M. K. 2001. The demographic basis of population regulation in Columbian ground squirrels. American Naturalist 158:236–247.
Feldhamer, G. A. 2002. Acorns and white-tailed deer: interrelationships in forest ecosystems. *In* Oak Forest Ecosystems: Ecology and Management for Wildlife, eds. W. J. McShea and W. M. Healy, pp. 215–223. John Hopkins University Press, Baltimore, Maryland.
Fitch, H. S., and Rainey, D. S. 1956. Ecological observations of the woodrat (*Neotoma floridana*). University of Kansas Publication of the Museum of Natural History 8:499–533.
Ford, W. M., Castleberry, S. B., Mengak, M. T., Rodrigue, J. L., Feller, D. J. and Russell, K. R. 2006. Persistence of Allegheny woodrats *Neotoma magister* across the mid-Atlantic Appalachian Highlands landscape, USA. Ecography 29:745–754.
Fowells, H. R. 1965. Silvics of the Forest Trees of the United States. Agricultural Handbook No. 271, USDA Forest Service, Washington, D.C.
Franklin, J. F., Berg, D. R., Thornburgh, D. A., and Tappeiner, J. C. 1997. Alternate silvicultural approaches to timber harvesting: variable retention harvest systems. *In* Creating a Forestry for the Twenty-First Century: The Science of Ecosystem Management, eds. K. A. Kohm and J. F. Franklin, pp. 111–139. Island Press, Washington, D.C.
Garrott, R. A., White, P. J., and Vanderbilt, C. A. 1993. Overabundance: an issue for conservation biologists? Conservation Biology 7:946–949.
Gilbert, B. S., and Krebs, C. J. 1981. Effects of extra food on *Peromyscus* and *Clethrionomys* populations in the southern Yukon. Oecologia 51:326–331.
Goodrich, L., Brittingham, M., Bishop, J., and Barber, P. 2006. An interagency-led effort: biodiversity conservation in Pennsylvania: a summary of the status of wildlife habitat and habitat threats statewide. *In* Biodiversity Conservation Handbook: State, Local, and Private Protection of Biological Diversity, eds. R. B. McKinstry, Jr., C. Ripp and E. Lisy, pp. 559–568. Environmental Law Institute, Washington, D.C.
Gorman, O. T. and Roth, R. R. 1989. Consequences of a temporally and spatially variable food supply for an unexploited gray squirrel *Sciurus carolinensis* population. American Midland Naturalist 121:41–60.
Hayden, A. 1988. Furbearer [raccoon] population and harvest on selected study areas. Job Code 06357-87, Final report to the Pennsylvania Game Commission, Harrisburg, Pennsylvania.
Hornsby, B. S., Ruiz, A. M., Castleberry, S. B., Castleberry, N. L., Ford, W. M., and Wood, P. B. 2005. Fall movements of Allegheny woodrats in harvested and intact stands in West Virginia. Northern Journal of Applied Forestry 22:281–284.
Hunter, M. L., Jr. 1996. Fundamentals of Conservation Biology. Blackwell Science, Cambridge, Massachusetts.
Irwin, H., Andrew, S., and Bouts, T. 2002. Return the Great Forest: A Conservation Vision for the Southern Appalachian Region. Southern Appalachian Forest Coalition, Asheville, North Carolina.
IUCN. 2001. IUCN Red List Categories and Criteria. Version 3.1. IUCN Species Survival Commission, Gland, Switzerland and Cambridge, United Kingdom.
Johnson, P. S. 1994. How to Manage Oak Forest for Acorn Production. Bulletin TB-NC-1, USDA Forest Service Northcentral Forest Experiment Station, Columbia, Missouri.

Latham, R. E., Beyea, J., Benner, J. M., Dunn, C. A., Fajvan, M. A., Freed, R. R., Grund, M., Horsley, S. B., Rhoades, A. F., and Shissler, B. P. 2005. Managing White-tailed Deer in Forest Habitat from an Ecosystem Perspective: Pennsylvania Case Study. Audubon Pennsylvania and Pennsylvania Habitat Alliance, Harrisburg, Pennsylvania.

LoGiudice, K. 1995. Control of *Baylisascaris procyonis* (Nematoda) in raccoons (*Procyon lotor*) through the use of antihelmintic baits. M.S. Thesis, Rutgers University, New Brunswick, New Jersey.

LoGiudice, K. 2000. *Baylisascaris procyonis* and the decline of the Allegheny woodrat (*Neotoma magister*). Ph.D. Dissertation, Rutgers University, New Brunswick, New Jersey.

LoGiudice, K. 2006. Toward a synthetic view of extinction: a history lesson from a North American rodent. Bioscience 56:687–693.

Lubow, B. C. 1996. Optimal translocation strategies for enhancing stochastic metapopulation viability. Ecological Applications 6:1268–1280.

MacKenzie, D. I., Nichols, J. D., Royle, J. A., Pollock, K. H., Bailey, L. L., and Hines, J. E. 2006. Occupancy Estimation and Modeling. Elsevier Academic Press, Burlington, Massachusetts.

McGowan, E. M. 1993. Experimental release and fate study of the Allegheny woodrat (*Neotoma magister*). Federal aid project report W-166-E-1, New York Department of Environmental Conservation, Delmar, New York.

McShea, W. J. 2000. The influence of acorn crops on annual variation in rodent and bird populations. Ecology 81:228–238.

McShea, W. J., and Healy, W. M. 2002. Oak Forest Ecosystems: Ecology and Management for Wildlife. John Hopkins University Press, Baltimore, Maryland.

McShea, W. J., and Schwede, G. 1993. Variable acorn crops: responses of white-tailed deer and other mast consumers. Journal of Mammalogy 74:999–1006.

Mehmood, S. R., and Zhang, D. 2001. Forest parcelization in the United States: a study of contributing factors. Journal of Forestry 99:30–34.

Mengak, M. T. 2002. Home range and movement of the Allegheny woodrat (*Neotoma magister*) in Virginia. Banisteria 19:3–8.

Myers, R. T. 1997. Microhabitat and ecology of the Allegheny woodrat in north-central West Virginia. M.S. Thesis, West Virginia University, Morgantown, West Virginia.

National Marine Fisheries Service. 2006. Interim endangered and threatened species recovery planning guidance. National Marine Fisheries Service, Washington, D.C. Available http://www.nmfs.noaa.gov/pr/recovery.

Nature Conservancy. 2001. Central Appalachian forest ecoregional plan. Nature Conservancy, Arlington, Virginia. Available http://conserveonline.org/docs/2005/03/CAP-plan˙2001.pdf

Nichols, J. D., Johnson, F. A., and Williams, B. K. 1995. Managing North American waterfowl in the face of uncertainty. Annual Review of Ecology and Systematics 26:177–199.

Noss, R. F. 1987. Protecting natural areas in fragmented landscapes. Natural Areas Journal 7:2–13.

Noss, R. F., and Cooperrider, A. Y. 1994. Saving Nature's Legacy. Island Press, Washington, D.C.

Noss, R. F., and Harris, L. D. 1986. Nodes, networks and MUM's: preserving diversity at all scales. Environmental Management 10:299–309.

Oehler, J. D., and Litvaitis, J. A. 1996. The role of spatial scale in understanding responses of medium-sized carnivores to forest fragmentation. Canadian Journal of Zoology 74:2070–2079.

Ostfeld, R. S., Jones, C. G., and Wolff, J. O. 1996. Of mice and mast: ecological connections in eastern deciduous forests. Bioscience 46:323–330.

Page, L. K., Gehrt, S. D., Titcombe, K. K., and Robinson, N. P. 2005. Measuring prevalence of raccoon roundworm (*Baylisascaris procyonis*): a comparison of common techniques. Wildlife Society Bulletin 33:1406–1412.

Pennsylvania Game Commission and Pennsylvania Fish and Boat Commission (PGC-PFBC). 2005. Pennsylvania wildlife action plan. Pennsylvania Game Commission and Pennsylvania Fish and Boat Commission, Harrisburg, Pennsylvania.

Poole, E. L. 1940. A life history sketch of the Allegheny woodrat. Journal of Mammalogy 21: 249–270.

Rodewald, A. D. 2003. Decline of oak forests and implications for forest wildlife conservation. Natural Areas Journal 23:368–371.

Rich, T. D., Beardmore, C. J., Berlanga, H., Blancher, P. J., Bradstreet, M. S. W., Butcher, G. S., Demarest, D. W., Dunn, E. H., Hunter, W. C., Iñigo-Elias, E. E., Kennedy, J. A., Martell, A. M., Panjabi, A. O., Pashley, D. N., Rosenberg, K. V., Rustay, C. M., Wendt, J. S., and Will, T. C. 2004. Partners in flight North American landbird conservation plan. Cornell Lab of Ornithology, Ithaca, New York.

Richmond, N. D., and Rosland, H. R. 1949. Mammal survey of northwestern Pennsylvania. Final report to the Pittman Robertson Project 20-R, Pennsylvania Game Commission, Harrisburg, Pennsylvania.

Rubino, D. L., and McCarthy, B. C. 2003. Composition and ecology of macrofungal and myxomycete communities on oak woody debris in a mixed-oak forest of Ohio. Canadian Journal of Forest Research 33:2151–2163.

Runge, J. P., Runge, M. C., and Nichols, J. D. 2006. The role of local populations within a landscape context: defining and classifying sources and sinks. American Naturalist 167:925–938.

Salafsky, N., Margoluis, R., and Redford, K. 2001. Adaptive Management: A Tool for Conservation Practitioners. Publication 112, Biodiversity Support Program, Washington, D.C.

Schneider, M. F. 2001. Habitat loss, fragmentation and predator impact: spatial implications for prey conservation. Journal of Applied Ecology 38:720–735.

Shaffer, M. L. 1981. Minimum population sizes for species conservation. Bioscience 31:131–134.

Soulé, M. E. 1990. The onslaught of alien species, and other challenges in the coming decades. Conservation Biology 4:233–239.

Tappeiner, J. C., Lavender, D., Walstad, J., Curtis, R. O., and DeBell, D. S. 1997. Silvicultural systems and regeneration methods: current practices and new alternatives. *In* Creating a Forestry for the Twenty-First Century: The Science of Ecosystem Management, eds. K. A. Kohm and J. F. Franklin, pp. 151–164. Island Press, Washington, D.C.

Tilman, D., May, R. M., Lehman, C. L., and Nowak, M. A. 1994. Habitat destruction and the extinction debt. Nature 371:65–66.

Topping, M. G., and Millar, J. S. 1996. Home range size of bushy-tailed woodrats, *Neotoma cinerea*, in Southwestern Alberta. Canadian Field-Naturalist 110:351–353.

Virginia Department of Forestry. 2006. State of the forest report. Virginia Department of Forestry, Charlottesville, Virginia.

Wilcove, D. S. 1987. From fragmentation to extinction. Natural Areas Journal 7:23–29.

Wilcox, B. A., and Murphy, D. D. 1985. Conservation strategy: the effects of fragmentation on extinction. American Naturalist 125:879–887.

Wink, J., Senner, S. E., and Goodrich, L. J. 1987. Food habits of great horned owls in Pennsylvania. Proceedings of the Pennsylvania Academy of Sciences 61:133–137.

Wright, J. 1998. Radiotelemetry study of experimental recolonization of an Allegheny woodrat (*Neotoma magister*) population. Final report to the Wild Resource Conservation Fund, Harrisburg, Pennsylvania.

Appendix: Guidelines for Identifying Priority Habitat Sites for Protection

The protection and management of habitat sites will be an important component of any recovery or management plan for *N. magister*. We have identified the following characteristics of habitat sites or metapopulation areas indicating they should receive priority for protection or management actions:

(1) *Larger habitat sites*—Protect larger habitat sites because they usually support more activity centers. In Pennsylvania, we found larger sites are more likely to be occupied, which is not evidence of area sensitivity but simply indicates that larger sites are more likely to contain at least one occupied activity center.
(2) *Occupancy*—Habitat sites with a greater proportion of occupied activity centers may be indicative of greater reproductive rates and intact dispersal corridors.
(3) *Metapopulation size*—The greater number and density of habitat sites in a metapopulation area indicate the potential for population viability. If habitat sites are less than 2 km apart, the metapopulation has greater potential for successful dispersal among habitat sites.
(4) *Tree species composition*—A habitat site dominated by tree species that produce hard mast potentially has better food resources available.
(5) *Less fragmentation*—Habitat sites with large nonforest tracts nearby (>100 ha in size and <1 km away) are likely to be adversely affected by predation and other stressors.
(6) *Roundworm contamination*—Habitat sites, particularly those close to farmlands or near riparian habitats or wetlands, chosen for *N. magister* management (especially reintroduction) should be surveyed for the presence of raccoon roundworm (Page et al. 2005). Habitat sites contaminated with roundworm should be considered habitat sinks.
(7) *Land ownership*—Habitat sites on public land should receive higher priority for conservation because they are more likely to retain beneficial characteristics than sites in private ownership.

Chapter 9
Genetic Diversity of Woodrats: Implications for Conservation

Timothy J. Smyser and Olin E. Rhodes, Jr.

Introduction

It is clear that, in conjunction with environmental and demographic stochasticity, genetic factors pose a significant threat to the persistence of small populations (Lande 1988, Leberg 1990, Mills and Smouse 1994) such as those that comprise Allegheny woodrat (*Neotoma magister*) metapopulations. In this chapter we will (1) discuss genetic issues that confront small populations; (2) address why Allegheny woodrat populations may be at particular risk from deleterious genetic effects; (3) summarize previously conducted genetic analyses of Allegheny woodrats; and (4) explore the potential consequences of inbreeding depression to the population dynamics of Indiana's Allegheny woodrats.

Genetic Susceptibility of Small Populations

Deleterious genetic effects are manifested in small populations through two independent, yet complementary processes: genetic drift and inbreeding depression. Genetic drift is the stochastic fluctuation in allele frequencies from one generation to the next as a consequence of the random sampling of gametes during fertilization (Hedrick 2005). Repeated random sampling, causing shifts in gene frequencies over successive generations, ultimately will result in loss of alleles and the concomitant fixation of one allele at the various loci. Such allelic fixation inherently limits a population's ability to adapt to changing environmental conditions, and the fixation of deleterious alleles reduces individual and population-level fitness. The rate at which a population's gene frequencies change is inversely related to effective population size, such that the potential for change in gene frequencies increases as population size decreases (Chesser et al. 1993, Rhodes and Chesser 1994).

T.J. Smyser
Department of Forestry and Natural Resources, Purdue University, West Lafayette, IN 47906

J.D. Peles, J. Wright (eds.), *The Allegheny Woodrat*, DOI: 10.1007/978-0-387-36051-5_9,

Given that genetic drift is the result of random sampling, gene frequencies of isolated subpopulations should drift independently. Predictions based on theoretical systems with an infinite number of subpopulations indicate that, although genetic diversity will be lost within each single subpopulation, the full range of genetic diversity will be preserved collectively among the many subpopulations. Similar processes of independent genetic drift are predicted to occur in metapopulation systems with finite numbers of subpopulations.

Concurrent with the continual reduction of genetic diversity through genetic drift, negative consequences associated with the loss of genetic diversity may emerge through inbreeding depression—the reduction in fitness associated with increasing homozygosity of detrimental alleles. Inbreeding depression has been documented across an array of taxa in captive, wild, and laboratory populations (Frankham and Ralls 1998). The effects of inbreeding depression are not only manifested through the mating of closely related individuals, but similar effects arise in small populations under a breeding system with random mating or inbreeding avoidance as a result of the increasing relatedness among potential breeders over successive generations.

Because of their small population sizes (Wood Chapter 3, Mengak et al. Chapter 7), the subpopulations that comprise Allegheny woodrat metapopulations are particularly vulnerable to the effects of genetic drift and inbreeding depression. In such metapopulations, the movement of individuals among subpopulations is imperative to reintroduce alleles lost from a subpopulation's gene pool through drift and prevent the onset of inbreeding effects. If connectivity among subpopulations is limited by factors such as increased habitat fragmentation or reduced productivity within subpopulations, genetic drift will proceed unabated and inbreeding effects will arise, ultimately influencing population dynamics. Such effects were observed within the Glanville fritillary butterfly (*Melitaea cinxia*) metapopulation on the Åland islands in southwest Finland, in which subpopulations with higher levels of inbreeding had greater extinction risks than noninbred subpopulations (Saccheri et al. 1998).

Rather than appearing as congenital abnormalities or sweeping reproductive failures, inbreeding depression typically is expressed through a decrease in vital rates or other fitness measures, which in turn collectively influence population dynamics (Mills 1996). Specifically, assessments of inbreeding effects have revealed reductions in survival rate, reproductive success, and population persistence (Jimenez et al. 1994, Keller et al. 1994, and Saccheri et al. 1998). Additionally, the consequences of inbreeding often may be manifested only during times of environmental stress (Keller et al. 1994, Keller 1998). Such was the case with mass die-offs of song sparrows (*Melospiza melodia*) on Mandarte Island (Keller et al. 1994, Keller 1998) in which inbred individuals perished at a much greater rate than noninbred conspecifics during severe winter weather. Using an experimental approach, Jimenez et al. (1994) provided corroborative evidence for the importance of environmental stress in revealing inbreeding depression in a small mammal species. Although survival rates were similar for inbred ($f = 0.25$) and noninbred white-footed mice (*Peromyscus leucopus*) in the laboratory environment, differences in mortality rates increased dramatically between members of the two groups when released into the natural environment with weekly survivorship of inbred mice being only 56% of that of their noninbred counterparts.

Reductions in survival associated with inbreeding typically are expressed as lethal equivalents. Lethal equivalents are calculated by regressing survival rates against inbreeding coefficients (Frankham et al. 2004) with one lethal equivalent representing a group of detrimental alleles that, when homozygous, results in the death of the individual. Ralls et al. (1988) summarized lethal equivalent loads from seven species of rodents by evaluating the proportion of individuals that survived to the lesser of one-half the age at sexual maturity or 180 days. Lethal equivalent loads among the seven species of rodents ranged from –0.14 to 15.16 (median = 1.15; Ralls et al. 1988). Based on data reported in Ralls et al. (1988), captive individuals born to the union of full sibs ($f = 0.25$) experienced a 40% decline in survival. Although populations vary in their susceptibility to inbreeding effects because of variation in past effective population sizes, breeding structure, and history of genetic purges, Allegheny woodrats may show a similar deleterious response to the effects of inbreeding.

Given the small population sizes of woodrat subpopulations with their concomitant susceptibility to genetic drift, and the woodrat's limited dispersal capacity, inbreeding effects may accrue even under conditions of random mating or inbreeding avoidance. If inbreeding effects for woodrats are similar to those documented for other rodent species (Ralls et al. 1988, Jimenez et al. 1994), dramatic declines in survival likely would threaten the persistence of populations, particularly in environments characterized by increasing population pressure associated with anthropogenic change.

Genetics of Allegheny Woodrats

A recent trend of population decline has occurred throughout much of the Allegheny woodrat's distribution (Ford et al. 2006, LoGiudice 2006, Wright Chapter 1), with patch extirpation and increasing isolation accompanying these declining population trends (Hudson et al. 2006). Local distributions of *N. magister* are generally restricted by the presence of rocky habitats (Castleberry et al. 2006): habitats that may be disjunct or rare even in the absence of anthropogenic land use change. Given the patchy distribution of suitable rocky habitats, woodrat populations commonly appear to be organized as metapopulations within landscapes (Manjerovic 2004, Wood Chapter 3).

Theoretical genetic models indicate that when subpopulations are numerous, spatially structured subpopulations are well suited for retention of genetic diversity within the entire system (Chesser et al. 1993, 1996), as independent, random genetic drift allows components of genetic diversity to be preserved among the many isolated subpopulations. However, given a probability of subpopulation extinction, as found in metapopulations, the predicted ability for the system to retain genetic diversity is reduced greatly. Under such a scenario, the ability of the overall metapopulation to retain genetic diversity becomes a function of extinction rate, connectivity among subpopulations, number of subpopulations, and number of founders recolonizing vacant patches (Hedrick and Gilpin 1997). Thus, as extrinsic pressures on spatially structured Allegheny woodrat metapopulations increase (LoGiudice 2006), subpopulation extinction rates will increase and connectivity will decrease (Ford et al. 2006), likely reducing the ability of woodrat metapopulation systems to retain genetic diversity. By examining patterns of neutral genetic diversity across a landscape, rates of connectivity among woodrat populations and subpopulations as well as long-term patterns of population size can be quantified.

Mitochondrial DNA Studies

Mitochondrial DNA (mtDNA) often has been used to resolve ancient patterns of ancestry such as those between species, subspecies, or very divergent populations within species. Mitochondrial DNA is passed down directly from mother to offspring without male contributions. Reducing the effective population to represent only females allows haplotype (differing versions of mtDNA genes, analogous to alleles) frequencies to drift more rapidly than those of biparentally inherited nuclear DNA.

Initial assessments of genetic structure for *Neotoma* species used mtDNA to evaluate distinctions among species and to infer patterns of haplotypic diversity within Allegheny woodrats (Hayes and Harrison 1992). To characterize mtDNA diversity, Hayes and Harrison (1992) used restriction fragment length polymorphism analysis (RFLP), a technique which employs restriction enzymes isolated from bacteria to digest DNA. Diversity of mtDNA is then inferred based on the number and length of fragments produced through the enzymatic digestion. Hayes and Harrison (1992) observed a common haplotype throughout the range of the Allegheny woodrat while the occurrence of additional haplotypes was restricted to regions or single sampling locations. They attributed the observed pattern of haplotypic diversity to the recent (i.e., last 10,000 years) expansion of *N. magister* from a southern Appalachian refugium with the recession of the Wisconsin glacial ice sheets. The low haplotype diversity found throughout the species' range suggests that a limited number of individuals initiated recolonization following a population bottleneck during the glacial extreme.

Nuclear DNA Studies

Microsatellites are biparentally inherited, neutral markers that typically exhibit very high levels of polymorphism, suitable for the evaluation of contemporary patterns of population isolation and connectivity. Microsatellites are tandem repeats of DNA sequence motifs: typically 1–5 bases in length (Frankham et al. 2004). The high polymorphism observed in microsatellites results from mutations occurring as a consequence of slippage of the replicating frame during DNA replication, which produces variation in the number of repeated sequence motifs (Frankham et al. 2004). Microsatellites offer advantages over mtDNA markers in that their high polymorphism and biparental inheritance provide the resolution necessary to characterize fine-scale genetic structure, which often is beyond the discriminatory power of mtDNA data (Frankham et al. 2004).

Castleberry et al. (2002) used 11 polymorphic microsatellite markers to describe genetic diversity across the Allegheny woodrat's range and patterns of connectivity within local woodrat metapopulations. Individuals from nine woodrat populations—representing areas on both the periphery (New Jersey, North Carolina, and western Kentucky) and core (Virginia, West Virginia, and eastern Kentucky) of the species' distribution—were sampled. Significant genetic differentiation was detected among populations, as the overall F_{ST} value among the nine populations was 0.17, with pairwise values ranging from 0.05 to 0.46. Genetic differentiation followed a pattern of isolation by distance, with genetic differentiation increasing as a function of increasing geographic distance. Accordingly, the greatest genetic distance observed was between populations from New Jersey and northeastern Kentucky ($F_{ST} = 0.46$; Castleberry et al. 2002).

The small disjunct subpopulations, characteristic of Allegheny woodrat metapopulations, provide a context where genetic differentiation could occur at very fine spatial scales. Treating a subset of these nine woodrat populations as metapopulations and evaluating disjunct habitat patches independently, Castleberry et al. (2002) assessed patterns of fine-scale genetic structure in local woodrat assemblages. Subpopulations separated by only 2 km exhibited F_{ST} values as high as 0.24, leading to the conclusion that "limited effective dispersal [occurred] among subpopulations separated by <3 km" (Castleberry et al. 2002). Using data provided in Castleberry (2000), we evaluated patterns of isolation by distance within these same metapopulations by regressing measures of genetic differentiation ($F_{ST}/(1-F_{ST})$) against geographic distance (Rousset 1997). Our analysis focused solely on patterns of genetic differentiation within metapopulations and omitted relationships among metapopulations. Genetic structure of woodrat metapopulations demonstrated a strong pattern of isolation by distance ($p < 0.00001$) indicating that high levels of genetic differentiation occur even within metapopulations. Therefore, based on the results from Castleberry et al. (2002) and our reanalysis of their data, woodrat subpopulations appear to have very small effective population sizes with limited gene flow among subpopulations. This finding suggests that declining woodrat populations may be vulnerable to genetic effects such as rapid rates of genetic drift, resulting in reduced genetic diversity, ultimately culminating in negative fitness consequences.

Working within one West Virginia metapopulation previously studied by Castleberry et al. (2002), Manjerovic (2004) also identified fine-scale patterns of isolation by distance similar to those observed in our analysis. Interestingly, in Manjerovic's (2004) study, the two subpopulations with the lowest pairwise F_{ST} value were separated by the Cheat River. This finding is not entirely surprising given that Castleberry et al. (2002) also observed that genetic differentiation across rivers was not greater than expected based on Euclidian distance. Therefore, subpopulations may remain connected despite the fact that rivers appear to human observers to be geographic barriers to dispersal. While demographic data have not confirmed the movement of woodrats across the Cheat River (Manjerovic 2004), it is possible that woodrats may cross during low summer flows or winter ice-over events. Alternatively, the river indeed may pose a barrier to dispersal, although sufficient time following colonization may not have elapsed for gene frequencies to diverge (Castleberry et al. 2002).

Although patterns of isolation by distance were observed among the subpopulations within this West Virginia metapopulation, Manjerovic (2004) concluded that contemporary levels of dispersal among subpopulations were sufficient to sustain current levels of genetic diversity within subpopulations. Castleberry et al. (2002) echoed this conclusion but cautioned that the functional population unit pertinent to Allegheny woodrats may be below the subpopulation level as defined in these studies. Consequently, management plans for this species should recognize the interbreeding group of individuals associated with each rock outcrop as a subpopulation. With such small functional population units, the preservation of forest corridors is imperative for the persistence of woodrat populations at multiple spatial and biological scales: from the demes associated with rock outcroppings to metapopulations (Castleberry et al. 2002, Ford et al. 2006).

Levels of population genetic diversity frequently reflect geographic relationships within the species' range (Schwartz et al. 2003). A common pattern found across a wide array of taxa is that genetic diversity is highest within populations inhabiting the core of the species' range and lowest in populations on the periphery of the range (Schwartz et al. 2003). This pattern is explained by the fact that core populations have a greater potential to receive migrants than do peripheral populations, as immigrants into core populations could come from all directions. Alternatively, dispersal opportunities into peripheral populations are more limited, as peripheral populations only can receive individuals that are moving in a direction away from the core of the species' range (Channell and Lomolino 2000, Schwartz et al. 2003). Interestingly, patterns of genetic diversity in Allegheny woodrats did not fully support the expected relationship between core and peripheral populations (Castleberry et al. 2002). For example, across the range of the Allegheny woodrat, peripheral populations exhibited among the highest and lowest levels of heterozygosity and average numbers of alleles per locus (western Kentucky $H = 75.0\%$, $A = 5.3$; New Jersey $H = 28.7\%$, $A = 2.0$; Castleberry 2000, Castleberry et al. 2002). This observation can be explained by the Allegheny woodrat's limited dispersal capacity (Wood Chapter 3) and the low abundance associated with most extant subpopulations, which has allowed local issues related to past and contemporary isolation and population sizes to overwhelm typical patterns of genetic diversity based on geographic location within the species' range. The failure of geographic relationships to accurately predict genetic patterns indicates that the stochastic process of genetic drift is driving genetic parameters within woodrat populations.

Irrespective of the role of stochasticity in shaping the range-wide pattern of genetic diversity in Allegheny woodrats, the importance of genetic drift and isolation in declining populations will be magnified as extrinsic pressures on woodrat populations increase. The New Jersey woodrat population exemplifies this process as, following a recent range contraction (LoGiudice 1995), reduced genetic diversity threatens the persistence of this population. Management intervention, either through establishment of natural or artificial connectivity of the New Jersey woodrat population to more genetically robust populations, appears to be necessary to re-establish its genetic diversity and to provide for the long-term persistence of this population (Castleberry et al. 2002). A similar interaction of population pressures with an ensuing disruption of genetic processes also may threaten the persistence of the woodrat metapopulation in Indiana.

Genetics of Indiana's Allegheny Woodrat Population: A Case Study

Similar to other Allegheny woodrat populations throughout the species' range, those in Indiana have demonstrated a precipitous decline recently, despite legal protection as a state endangered species since 1984 (Johnson 2002). At the time of listing, little was known about local woodrat population trends, although the species was deemed at risk because of its restricted distribution and small population sizes. Concerted efforts were made in the early 1980s to map the distribution of woodrat populations within the state (Cudmore 1983). Although fossil evidence indicated that woodrats once were widespread in southern Indiana (ca. 145 km north of their current range; Richards 1987) their current distribution is largely restricted to limestone escarpments that border the Ohio River in extreme southern Indiana (Johnson 2002).

Indiana's Department of Natural Resources initiated an Allegheny woodrat population monitoring program in 1991, using live-trapping as a means of assessing population trends and patch occupancy (Mengak et al. Chapter 7). From its inception, the monitoring program documented a clear pattern of population decline with repeated patch extirpations. Of the 19 sites with documented woodrat occupation from 1980 to 1983, six sites were extirpated by 1992. Although Indiana's woodrat population appeared stable from 1991 through 1996, populations declined precipitously from 1996 to 2002. By 2002, population indices suggested that woodrat abundance had declined by 34% over the previous six years with an additional three of the remaining 13 patches having been extirpated (Mengak et al. Chapter 7). As of 2006, woodrat populations seem to have temporarily stabilized with some patches appearing to have been recolonized. However, many of the extant woodrat subpopulations remain on the brink of extirpation with persistence threatened by very low abundance and skewed sex ratios.

Several hypotheses have been proposed to explain the declines in Indiana's Allegheny woodrat populations: (1) habitat fragmentation; (2) reduction in food resources; and (3) increased fatal exposure to the parasitic raccoon roundworm (*Baylisascaris procyonis*; LoGiudice 2006, Chapter 2). Additionally, small population size and geographic isolation make Indiana's woodrat populations vulnerable to negative effects associated with losses of genetic diversity and inbreeding depression.

Knowledge gained through intensive population monitoring efforts exerted over the past 25 years provides an ideal context from which to evaluate the genetic patterns of Indiana's declining woodrat metapopulation. Thus, studies were initiated in 2005 to characterize patterns of genetic diversity and connectivity within and among Indiana's remaining woodrat subpopulations in order to assess genetic risks to population persistence. By employing the same suite of 11 microsatellite markers used by Castleberry et al. (2002), we were able to directly compare levels of genetic diversity to other populations surveyed throughout the species' range. Three distinct genetic units were identified within Indiana (East, Central, and West; Fig. 9.1) with levels of geographic isolation similar to those observed among subpopulations comprising the metapopulations in Castleberry et al.'s (2002) analysis. Relative to other populations throughout the species' range, much of the potential genetic diversity already has been lost from Indiana's woodrat metapopulation. Observed heterozygosity values and average numbers of alleles per locus were East 5.8%, 1.27; Central 36.2%, 2.45; and West 14.2%, 1.27; in contrast to the range-wide average of 61.8% and 5.4 observed by Castleberry et al. (2002). Undoubtedly, this lack of genetic diversity will jeopardize the future persistence of Indiana's woodrat populations.

In addition to evidence of low genetic diversity, there appears to be little prospect for the natural restoration of genetic diversity through exchange of individuals among populations; our data provide no evidence for the presence of migrants or the descendants of migrants within any of the Indiana populations. Despite relatively close proximity of remnant woodrat populations (<10 km), there appears to be sufficient distance or landscape resistance between sites to prevent the movement of individuals. Genetic distances expressed as pairwise F_{ST} values among the three extant woodrat populations in Indiana exceeded those observed between populations

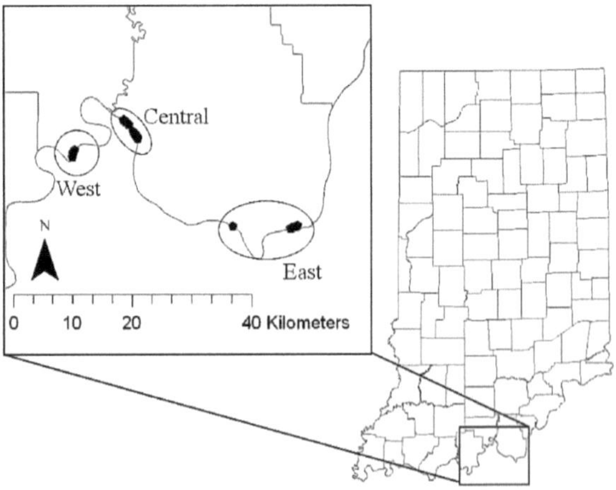

Fig. 9.1 Three Allegheny woodrat (*Neotoma magister*) populations in Indiana. The Central population is the largest and most genetically diverse while the peripheral East and West populations are genetically depauperate

Table 9.1 Pairwise values for genetic differentiation (F_{ST}—below the diagonal) and Euclidean distance in km (above the diagonal) for three genetically distinct populations of Allegheny woodrats in Indiana

	East	Central	West
East	–	28.5	34.2
Central	0.47	–	9.9
West	0.78	0.41	–

at the extremes of the species' range (Table 9.1; Castleberry et al. 2002). Without the movement of individuals among populations, the very low abundances associated with Indiana's woodrat populations have allowed gene frequencies to drift rapidly and genetic diversity to be lost within the three remaining breeding colonies.

With significantly depressed levels of genetic diversity and little prospect for natural restoration, the Indiana woodrat population appears to be entering an extinction vortex, a positive feedback system in which poor demographic performance and genetic processes reinforce one another, further depressing abundance and elevating the risk of extirpation (Gilpin and Soulé 1986, Frankham et al. 2004). With little probability for natural connectivity among Indiana's remnant populations, artificial supplementation appears to be the only available tool to restore genetic diversity within these populations. The introduction of a limited number of individuals can improve the fitness of genetically depauperate populations (Heschel and Paige 1995, Tallmon et al. 2004). Examples of such a genetic rescue effect can be drawn from a number of taxa including mustard (*Brassica campestris*; Newman and Tallmon 2001), adders (*Vipera berus*; Madsen et al. 1996, 1999), and prairie chickens (*Tympanuchus cupido pinnatus*; Bouzat et al. 1998a,b, Westemeier et al. 1998).

Given that a lack of genetic diversity threatens the immediate persistence of some of Indiana's woodrat populations, translocation of genetically diverse individuals into the East and West populations appears to be a prudent management strategy, although translocation is a management tool that should be used only in appropriate circumstances and after careful consideration (Serfass Chapter 10). Imprudent movement of individuals among populations in the name of genetic restoration can contaminate relic gene pools and alter the evolutionary trajectory of local populations (Moritz 1999, Myers and Knoll 2001).

Given that patterns of connectivity and selection operate differentially across the populations that comprise a species, emphasis in translocations is given to the use of local stocks to safeguard the evolutionary process. Under appropriate circumstances, the translocation of individuals between geographically proximate populations presents a number of benefits over translocations involving geographically distant conspecifics. Adjacent populations typically exhibit similar histories of genetic exchange and selective pressures. The movement of individuals among adjacent populations most readily mimics the natural process of gene flow in wild populations and maintains the genetic integrity of supplemented populations. Translocation of individuals among adjacent populations also minimizes risks associated with outbreeding depression. Outbreeding depression occurs through the introduction of individuals who are behaviorally or morphologically maladapted for the unique selective pressures of their new environment. Additional outbreeding effects can occur through underlying genetic differences that may arise without morphological manifestations such as the disassociation of beneficially linked gene complexes (Moritz 1999, Storfer 1999, Tallmon et al. 2004). For example, populations that occur within similar environments but have been historically isolated may have evolved unique gene complexes in response to similar selective pressures. Mixing such populations would lead to the uncoupling of the genetic linkage groups responsible for these unique adaptations and could lead to outbreeding depression in the admixed population. Finally, using adjacent populations in translocations minimizes the risk of introducing new diseases to naïve populations (Storfer 1999).

While a number of factors encourage the use of local genetic stocks for translocation efforts, a lack of genetic diversity or abundance may preclude their use in conservation programs. Reintroductions are inherently founder events in that a limited number of individuals are used to establish each new population. The extent to which founding individuals capture the genetic diversity present within a source population is limited by the number of founding individuals and the bias associated with how founding individuals were sampled from their native population. Specifically, the use of closely related individuals in a reintroduction can greatly reduce the diversity of genetic information captured within a fixed number of founders. Upon reintroduction, the amount of genetic diversity perpetuated from the founders depends upon survival/dispersal rates, proportion of founders which successfully breed, evenness of sex-specific reproductive variance, and population growth rate. With these post-reintroduction constraints further reducing the potential amount of genetic diversity that can be established by a reintroduction, it is imperative to select a genetically diverse source population to provide for the long-term adaptability of a reintroduced population (Moritz 1999). If local sources are genetically depauperate, managers must weigh the cost associated with reduced genetic diversity against the benefits of geographic proximity in selecting the most appropriate reintroduction strategy for a species.

In the context of Indiana's woodrat populations, two (East and West) of the three extant populations have lost much of their genetic diversity, and a genetic restoration is needed to provide for the short-term persistence of these populations. To identify the most appropriate conservation strategy for these populations, two questions must be addressed. Is there sufficient allelic richness retained within the Indiana woodrat populations to restore genetic diversity to the East and West populations through the movement of individuals among adjacent populations? Do the Indiana woodrat populations represent a unique evolutionary trajectory that might be contaminated or swamped through the introduction of foreign individuals?

For Indiana's supplementation efforts, the answer to the first question requires an assessment of the genetic diversity retained among Indiana's woodrat populations relative to levels of genetic diversity observed for populations across the species' range. Castleberry et al. (2002) reported a mean heterozygosity level of 61.8% with an average of 5.4 alleles per locus. These mean values included New Jersey, a woodrat population of conservation concern that, much like Indiana's woodrat populations, demonstrated reduced levels of genetic diversity ($H = 28.7\%$, $A = 2.0$).

Mean values of H and A increased to 65.9% and 5.8, respectively when considering only the eight more robust populations included in the range-wide survey (Castleberry 2000). The Central population, Indiana's most genetically diverse, exhibits heterozygosity levels at approximately half of those observed across the species' range ($H = 36.2\%$), and substantially reduced numbers of alleles per locus ($A = 2.5$). Thus, from a strictly genetic standpoint, the Central population would be a poor source population for supplementation and genetic restoration of the genetically depauperate East and West colonies.

If Indiana woodrat populations demonstrate a long history of isolation from the core of the species' range with a unique suite of selective pressures, these populations may represent a unique evolutionary trajectory. Furthermore, if isolation and selective pressures experienced by Indiana's woodrat population have resulted in significant genetic divergence of this population from other Allegheny woodrat populations in the species' range, then emphasis should be given to preserving the unique evolutionary trajectory of the Indiana woodrat population. Under such a scenario, appropriate source populations would be limited to adjacent rather than disjunct populations. In evaluating the potential for unique evolutionary trajectories to arise within woodrats, patterns of mtDNA diversity indicate Allegheny woodrats emerged from a single glacial refugium within the last 10,000 years with limited haplotype divergence (Hayes and Harrison 1992). As habitats were colonized during the retreat of the glacial ice sheet, patterns of isolation developed as a function of geographic distance (Castleberry et al. 2002). Given the recent time-frame in which Allegheny woodrats populations are thought to have descended from a common ancestral source, sufficient time likely has not yet passed for Indiana's Allegheny woodrat or other populations throughout the range to diverge on unique evolutionary trajectories.

The conclusion of limited evolutionary trajectory divergence also is corroborated by the rarity of private microsatellite alleles among woodrat populations throughout their range (Castleberry et al. 2002) and within Indiana. If Indiana woodrat populations had a long history of isolation, then we might expect to find a number of private alleles, even following substantial levels of genetic drift. Alternatively, comparison of the distribution of alleles between Indiana's woodrat population and the nine populations evaluated by Castleberry (2000) indicated that the majority of alleles observed in Indiana are contained in populations elsewhere throughout the species' range. Based on these lines of evidence, appropriate source populations for translocation into Indiana would be those populations exhibiting high levels of genetic diversity, experiencing similar environmental conditions, and that are most proximate to Indiana.

Conservation and Management of Declining Species

Because of their strict habitat affinities, Allegheny woodrat populations are typically organized as spatially structured metapopulations. Compared with other declining metapopulations, the characteristically small, disjunct breeding groups of Allegheny woodrats are particularly susceptible to the effects of genetic drift and inbreeding depression. When extrinsic pressures disrupt natural patterns of connectivity among networks of subpopulations, genetic catastrophe can arise throughout the entire metapopulation system.

Genetic assessments of Allegheny woodrat populations indicate that losses of genetic diversity threaten the persistence of some populations. Following a recent trend of range contraction and patch extinction, the remnant populations in New Jersey and Indiana show severely depressed levels of allelic richness and heterozygosity. With little probability of genetic restoration through immigration from adjacent populations, these populations' lack of genetic diversity likely will constrain their ability to adapt to changing environmental pressures and ultimately result in their extirpation. Clearly, the current lack of genetic diversity in these and potentially other isolated woodrat populations (e.g., Edge of Appalachia Preserve in Ohio) will hinder recovery efforts and must be mitigated for successful conservation efforts. This pattern of population isolation, followed by loss of genetic diversity, ultimately threatening population persistence, may reappear throughout small populations across an array of taxa under the ever-increasing pressure of habitat fragmentation.

Metapopulations are uniquely sensitive to the effects of fragmentation in that the loss of interconnecting habitat corridors as a result of changing land use patterns can dramatically alter genetic processes. Lessons from previous genetic surveys of woodrat populations have indicated that "the retention of forested dispersal corridors between proximate rock outcrops is critical to maintain current patterns of gene flown and genetic structure within subpopulations" (Castleberry et al. 2002). Manjerovic (2004) concluded that current levels of gene flow within a West Virginia metapopulation were sufficient to overcome inbreeding effects, but inbreeding effects would arise if levels of connectivity were reduced. Thus, although suitable woodrat habitat may be buffered from the immediate effects of habitat destruction in some locations (Balcom and Yahner 1996), fragmentation of adjacent forest corridors can dramatically alter the probability of patch persistence (Verboom and van Apeldoorn 1990) and the ability of these systems to retain genetic diversity.

Conservation of Allegheny woodrat metapopulations is a unique case in that not only are historic patterns of subpopulation connectivity being disrupted by the ubiquitous effects of habitat fragmentation, but new population pressures such as declines in food availability and increased mortality also have the capacity to reduce population productivity, and thus, the potential for gene flow through dispersal. This is particularly true if the production of dispersing woodrats is a function of habitat saturation, in that individuals only disperse when all locally available habitats are occupied. A reduction in the dispersing cohort would lead not only to decreased levels of gene flow among subpopulations but also would reduce the potential for recolonization of unoccupied patches, ultimately diminishing occupancy rates of intermediate habitats within metapopulation complexes and steadily decreasing rates of connectivity. Additionally, increasingly isolated populations may begin to manifest inbreeding effects which would reduce population recruitment, further decreasing the numbers of potential dispersing individuals and concomitantly increasing population isolation.

In recognition of the fact that woodrat populations now persist in fragmented landscapes with new anthropogenically introduced population pressures such as chestnut blight (*Cryphonectria parasitica*), gypsy moth (*Lymantria dispar*) infestation, and exposure to *B. procyonis*, managers should continue seeking ways to maintain connectivity among disjunct populations. If connectivity within a metapopulation has been severed by human activities, maintaining genetic diversity and population fitness within the metapopulation may require the proactive movement of individuals among subpopulations in a way that simulates natural patterns of connectivity. The proactive movement of a limited number of individuals within fragmented Allegheny woodrat metapopulations will allow for the preservation of levels of genetic diversity and fitness without risking the potentially deleterious effects of using more remote populations for the restoration of completely depleted genetic diversity.

Given the naturally disjunct distribution of their populations, Allegheny woodrats serve as a model species for research pertaining to the ecology, management, and conservation of metapopulations. As native habitats are lost and become increasingly fragmented, threats to the persistence of species due to the isolation and subsequent loss of genetic polymorphisms within and among population components will become of paramount importance to the conservation of species inhabiting human dominated landscapes. Lessons learned through research on and conservation efforts for the Allegheny woodrat will enable natural resource managers to more effectively manage and conserve species inhabiting naturally disjunct and anthropogenically fragmented environments.

References

Balcom, B. J., and Yahner, R. H. 1996. Microhabitat and landscape characteristics associated with the threatened Allegheny woodrat. Conservation Biology 10:515–525.

Bouzat, J. L., Cheng, H. H., Lewin, H. A., Westemeier, R. L., Brawn, J. D., and Paige, K. N. 1998a. Genetic evaluation of a demographic bottleneck in the greater prairie chicken. Conservation Biology 12:836–843.

Bouzat, J. L., Lewin, H. A., and Paige, K. N. 1998b. The ghost of genetic diversity past: historical DNA analysis of the greater prairie chicken. American Naturalist 152:1–6.

Castleberry, S. B. 2000. Conservation and management of Allegheny woodrats in the central Appalachians. Ph.D. Dissertation, University of West Virginia, Morgantown, West Virginia.

Castleberry, S. B., King, T. L., Wood, P. B., and Ford, W. M. 2002. Microsatellite DNA analysis of population structure in Allegheny woodrats (*Neotoma magister*). Journal of Mammalogy 83:1058–1070.

Castleberry, S. B., Mengak, M. T., and Ford, W. M. 2006. *Neotoma magister*. Mammalian Species 789:1–5.

Channell, R., and Lomolino, M. V. 2000. Dynamic biogeography and conservation of endangered species. Nature 403:84–86.

Chesser, R. K., Rhodes, O. E., Jr., and Smith, M. H. 1996. Conservation. *In* Population Dynamics in Ecological Space and Time, eds. O. E. Rhodes, Jr., R. K. Chesser and M. H. Smith, pp. 237–252. University of Chicago Press, Chicago, Illinois.

Chesser, R. K., Rhodes, O. E., Jr., Sugg, D. W., and Schnabel, A. 1993. Effective sizes for subdivided populations. Genetics 135:1221–1232.

Cudmore, W. W. 1983. The distribution and ecology of the Eastern woodrat, *Neotoma floridana*, in Indiana. Ph.D. Dissertation, Indiana State University, Terre Haute, Indiana.

Ford, W. M., Castleberry, S. B., Mengak, M. T., Rodrigue, J. L., Feller, D. J., and Russell, K. R. 2006. Persistence of Allegheny woodrats *Neotoma magister* across the mid-Atlantic Appalachian Highlands landscape, USA. Ecography 29:745–754.

Frankham, R., Ballou, J. D., and Briscoe, D. A. 2004. Introduction to Conservation Genetics. Cambridge University Press, Cambridge, United Kingdom.

Frankham, R., and Ralls, K. 1998. Inbreeding leads to extinction. Nature 392:441–442.

Gilpin, M. E., and Soulé, M. E. 1986. Minimum viable populations: the processes of species extinctions. *In* Conservation Biology: The Science of Scarcity and Diversity, ed. M. E. Soulé, pp. 19–34. Sinauer Associates Incorporated, Sunderland, Massachusetts.

Hayes, J. P., and Harrison, R. G. 1992. Variation in mitochondrial-DNA and the biogeographic history of woodrats (*Neotoma*) of the Eastern United-States. Systematic Biology 41:331–344.

Hedrick, P. W. 2005. Genetics of Populations. Jones and Bartlett, Boston, Massachusetts.

Hedrick, P. W., and Gilpin, M. E. 1997. Effective genetic size of a metapopulation. *In* Metapopulation Biology: Ecology, Genetics, and Conservation, eds. I. A. Hanski and M. E. Gilpin, pp. 166–182. Academic Press, Boston, Massachusetts.

Heschel, M. S., and Paige, K. N. 1995. Inbreeding depression, environmental-stress, and population-size variation in Scarlet-Gilia (*Ipomopsis aggregata*). Conservation Biology 9:126–133.

Hudson, C. M., Johnson, S. A., Holbrook, A. L., Walker, H. D., and Walker, R. M. 2006. 2005 Allegheny woodrat monitoring program. Indiana Department of Natural Resources, Bloomington, Indiana.

Jimenez, J. A., Hughes, K. A., Alaks, G., Graham, L., and Lacy, R. C. 1994. An experimental study of inbreeding depression in a natural habitat. Science 266:271–273.

Johnson, S. A. 2002. Reassessment of the Allegheny woodrat (*Neotoma magister*) in Indiana. Proceedings of the Indiana Academy of Science 1:56–66.

Keller, L. F. 1998. Inbreeding and its fitness effects in an insular population of song sparrows (*Melospiza melodia*). Evolution 52:240–250.

Keller, L. F., Arcese, P., Smith, J. N. M., Hochachka, W. M., and Stearns, S. C. 1994. Selection against inbred song sparrows during a natural population bottleneck. Nature 372: 356–357.

Lande, R. 1988. Genetics and demography in biological conservation. Science 241:1455–1460.

Leberg, P. L. 1990. Influence of genetic variability on population growth: implications for conservation. Journal of Fish Biology 37:193–195.

LoGiudice, K. 1995. Control of *Baylisascaris procyonis* (Nematoda) in raccoons (*Procyon lotor*) through the use of anthelmintic baits: a potential method for reducing mortality in the Allegheny woodrat (*Neotoma floridana magister*). M.S. Thesis, Rutgers University, New Brunswick, New Jersey.

LoGiudice, K. 2006. Toward a synthetic view of extinction: a history lesson from a North American rodent. Bioscience 56:687–693.

Madsen, T., Shine, R., Olsson, M., and Wittzell, H. 1999. Restoration of an inbred adder population. Nature 402:34–35.

Madsen, T., Stille, B., and Shine, R. 1996. Inbreeding depression in an isolated population of adders *Vipera berus*. Biological Conservation 75:113–118.

Manjerovic, M. E. 2004. Demography and genetic structure of an Allegheny woodrat population in northcentral West Virginia. M.S. Thesis, West Virginia University, Morgantown, West Virginia.

Mills, L. S. 1996. Cheetah extinction: Genetics or extrinsic factors? Conservation Biology 10:315.

Mills, L. S., and Smouse, P. E. 1994. Demographic consequences of inbreeding in remnant populations. American Naturalist 144:412–431.

Moritz, C. 1999. Conservation units and translocations: strategies for conserving evolutionary processes. Hereditas 130:217–228.

Myers, N., and Knoll, A. H. 2001. The biotic crisis and the future of evolution. Proceedings of the National Academy of Sciences of the United States of America 98:5389–5392.

Newman, D., and Tallmon, D. A. 2001. Experimental evidence for beneficial fitness effects of gene flow in recently isolated populations. Conservation Biology 15:1054–1063.

Ralls, K., Ballou, J. D., and Templeton, A. 1988. Estimates of lethal equivalents and cost of inbreeding in mammals. Conservation Biology 2:185–193.

Rhodes, O. E., Jr., and Chesser, R. K. 1994. Genetic concepts for habitat conservation: the transfer and maintenance of genetic variation. Landscape and Urban Planning 28:55–62.

Richards, R. L. 1987. The Quaternary distribution of the eastern woodrat, *Neotoma floridana*, in Indiana. Proceedings of the Indiana Academy of Science 96:513–521.

Rousset, F. 1997. Genetic differentiation and estimation of gene flow from *F*-statistics under isolation by distance. Genetics 145:1219–1228.

Saccheri, I., Kuussaari, M., Kankare, M., Vikman, P., Fortelius, W., and Hanski, I. 1998. Inbreeding and extinction in a butterfly metapopulation. Nature 392:491–494.

Schwartz, M. K., Mills, L. S., Ortega, Y., Ruggiero, L. F., and Allendorf, F. W. 2003. Landscape location affects genetic variation of Canada lynx (*Lynx canadensis*). Molecular Ecology 12:1807–1816.

Storfer, A. 1999. Gene flow and endangered species translocations: a topic revisited. Biological Conservation 87:173–180.

Tallmon, D. A., Luikart, G., and Waples, R. S. 2004. The alluring simplicity and complex reality of genetic rescue. Trends in Ecology and Evolution 19:489–496.

Verboom, B., and van Apeldoorn, R. 1990. Effects of habitat fragmentation on the red squirrel, *Sciurus vulgaris* L. Landscape Ecology 4:171–176.

Westemeier, R. L., Brawn, J. D., Simpson, S. A., Esker, T. L., Jansen, R. W., Walk, J. W., Kershner, E. L., Bouzat, J. L., and Paige, K. N. 1998. Tracking the long-term decline and recovery of an isolated population. Science 282:1695–1698.

Chapter 10
Reintroduction of Woodrats: Concepts and Applications

Thomas L. Serfass

Introduction

The movement of wild animals for restoration purposes has become a necessary and more common aspect of modern conservation practices. The International Union for the Conservation of Nature and Natural Resources (IUCN 1998) defines *translocation* as the movement of living organisms from one area to another. *Reintroductions* and *restockings* (hereafter *supplementations*) are forms of translocations (Griffin et al. 1989, Kleiman 1989, IUCN 1998). A reintroduction is the intentional release of an organism into a portion of its historic range to restore populations extirpated because of human activities. In contrast, supplementation is the release of an organism to parts of its historic range where a remnant population of conspecifics exists and is usually undertaken to provide genetic or numeric enrichment to small, isolated populations. For these reasons post-reintroduction supplementations may be a planned component of reintroduction projects or an adaptive component determined appropriate from ongoing assessments and perceived benefits to a reintroduced population. Supplementations generally are recommended to be avoided when the small, remnant populations are the result of deteriorating habitat conditions. Similarly, post-reintroduction supplementations should not be undertaken if monitoring of the reintroduced populations reveals unforeseen conditions at release sites that may preclude the re-establishment of viable populations, regardless of the addition of new individuals.

Early reintroduction projects in the United States were undertaken primarily to restore wildlife that had been severely depleted during the 1800s through the early 1900s by unregulated hunting and trapping. These projects focused on species considered economically or recreationally important such as beaver (*Castor canadensis*), white-tailed deer (*Odocoileus virginianus*), and wild turkeys (*Meleagris gallopavo*). The attention given to restoring these species is understandable

T.L. Serfass
Department of Biology, Frostburg State University, Frostburg, MD 21532

J.D. Peles, J. Wright (eds.), *The Allegheny Woodrat*,
DOI: 10.1007/978-0-387-36051-5_10,

given that the origins and original missions of state wildlife agencies primarily were embedded in a need to reverse declines in wildlife populations caused by overharvest (Trefethen 1975). Gradually, however, state wildlife agencies have been adopting a more holistic approach as demonstrated by increasing attention to the conservation of both game and nongame (those not traditionally harvested) wildlife. The remedies for reversing declines in populations of nongame species and some game species typically are much more complicated to identify and address than those related to harvest management. Landscape-level perturbations associated with ongoing growth of human populations have continuously evolved into new, more insidious threats to wildlife (e.g., fragmentation of forested landscapes and pollution of aquatic habitats). Negative consequences of extensive landscape-level changes in habitat conditions are imparted on a greater variety of species than those that were imposed by unregulated harvest, and have contributed to an expanding number of species, both game and nongame, requiring conservation attention—in some cases warranting the implementation of reintroduction projects.

During the last 30 years, numerous successful wildlife reintroduction projects—often involving charismatic species (i.e., species with traits that generate high levels of public attention)—have been initiated in the United States. Examples include American martens (*Martes americana*), bald eagles (*Haliaeetus leucocephalus*), black-footed ferrets (*Mustela nigripes*), California condor (*Gymnogyps californianus*), fishers (*Martes pennanti*), gray wolves (*Canis lupus*), ospreys (*Pandion haliaetus*), peregrine falcons (*Falco peregrinus*), and river otters (*Lontra canadensis*). Nongame species have been well represented among these more recent reintroduction projects, but typically involved larger, better known species—those that generally can be characterized biologically as having relatively low levels of reproduction and high rates of survival (*K*-selected species).

Substantial population declines experienced by many species of *Neotoma* have prompted increasing attention for development of active conservation plans for the species of greatest conservation concern (e.g., Ford et al. 2006, McCleery et al. 2006, Feldhamer and Poole Chapter 11). Central among considerations for more active management is the conservation value that can be derived by initiating reintroduction projects, particularly for the Allegheny woodrat (*Neotoma magister*). Unfortunately, as is the case for most *r*-selected species, there are no comprehensive procedures and only a few published case studies with direct application for guiding the overall process of reintroducing woodrats. In this chapter, I review some past unsuccessful Allegheny woodrat reintroduction efforts and provide a theoretical and practical framework for planning more successful programs for this and similar species.

Woodrat reintroduction projects have taken place in four states and all have involved the Allegheny woodrat, where it had become extirpated in northern portions of the range. These reintroductions included the release of 44 woodrats from Ohio and Kentucky at a site in Ohio during 1983 and 1984 (Schlie 1985, Schlie and Bookhout 1985); 29 woodrats from West Virginia at two sites in New York during 1991 (McGowan 1993); 42 woodrats from Kentucky and West Virginia at four sites in New Jersey during 1996 (LoGiudice 2003); and 10 woodrats from one site to another in Pennsylvania (source population was about 20 km from the release site) during 1997 (Wright 1998). Radiotelemetry studies were conducted to monitor fates of woodrats at all reintroduction sites, with outcomes indicating that the reintroductions did not contribute to the establishment of self-sustaining populations because of low survival and reproduction. The failures of the aforementioned projects demonstrate the challenges involved in restoring mammals with *r*-selected characteristics (in this case Allegheny woodrats) and the need to further develop and refine protocols for reintroduction projects involving this group of animals.

The IUCN's Reintroduction Specialist Group (RSG) provides excellent, general guidelines for developing and implementing reintroduction projects (IUCN 1998). Nonetheless, the various components of a project ideally should be adapted for the particular needs of the species being reintroduced—although similar approaches typically can be applied among species possessing comparable life-history traits. The proportionally large variety of *K*-selected species reintroduced has contributed to the establishment of reasonably well-defined protocols for guiding approaches in future projects involving similarly adapted species (Fritts et al. 1997, Griffin et al. 1989, Kleiman 1989, Serfass et al. 2003). In contrast, few reintroduction projects have been conducted involving small, short-lived species with high reproductive and mortality rates (*r*-selected). Consequently, there are few examples available to guide reintroductions of *r*-selected species (Jordan 2003).

The reintroduction process would benefit both small and larger mammals if the general guidelines prescribed by the IUCN-RSG are followed. However, the reintroduction of small mammals poses a variety of different challenges—many related to their relatively low survivorship in comparison to large mammals. The additional attention now being devoted to nongame species has contributed to the awareness that an increasing number of small mammal species are becoming of conservation concern and may require more active management, sometimes including the implementation of reintroduction projects, to ensure maintenance of diverse, viable populations.

The objective of any reintroduction project should be to establish a biologically meaningful population that contributes to the long-term conservation goals established for the species. The ensuing discussion focuses on developing a general set of procedures that can be applied for guiding and enhancing the process of reintroducing woodrats. The approach generally follows those prescribed by the IUCN-RSG, but applies added emphasis on specific considerations that should be applied for the reintroduction of *r*-selected species and, when possible, integrating appropriate conservation and biological information pertaining specifically to woodrats.

Preliminary Assessments

Reintroduction projects are inherently complex, involving an array of biological and social issues. Consequently, a variety of preliminary assessments should be undertaken as a basis for establishing justifications and protocols prior to initiating the reintroduction process. Presumably, for a species to be regarded as a candidate for reintroduction, evidence exists suggesting that its current range has undergone substantial declines in comparison to the historic range. Nonetheless, foremost among the preliminary assessments is a thorough review of documentation pertaining to the historic and current distributions of the species.

In the case of the Allegheny woodrat, declines in portions of the range are now well-documented, but initial awareness of the decline largely was based on anecdotal observations (Wright Chapter 1). For cryptic species like woodrats, the most definitive evidence of historic range is specimen records in museum collections (Wright Chapter 1). Unfortunately, for many nongame species, data on current distribution are not routinely collected to compare with historic museum specimens, so even this preliminary step may require launching a field survey of historic sites. Surveys of this type can be a substantial undertaking (Mengak et al. Chapter 7) and have yet to be accomplished for all states within the historic range of the Allegheny woodrat (Wright Chapter 1). If the review supports the presumption that the species' distribution has declined to an extent warranting conservation concern, then a feasibility study should be undertaken to more formally assess justifications, implications, and logistical and biological considerations for implementing a reintroduction project.

The primary aim of a feasibility study should be to put forth clear and concise rationales, based on a thorough assessment of the best available information, for use by decision makers in determining the justification and conservation value of implementing a reintroduction project (IUCN 1998). The feasibility study should also elucidate important procedural components of a reintroduction project based on biological requirements and ecological relationships of the species being assessed. Finally, the assessment must identify resources (e.g., cost and expertise) anticipated to be necessary for proper support of all phases of the reintroduction project.

The outcome (final product) of a well-conceived feasibility study should provide:

(1) A complete review of the literature pertaining to the basic biology, taxonomy, ecological associations, and techniques applicable to the conservation of the species (e.g., best methodologies for capture and handling). An outcome of this process should be the compilation of a well-organized library of pertinent readings and an associated annotated bibliography. A recent literature review for the Allegheny woodrat (Castleberry et al. 2006) and the chapters of this book provide a comprehensive, up-to-date literature survey and will serve as an excellent basis for establishing such a library.

(2) A comprehensive reassessment of existing information about the distribution of wild populations for use in determining if additional surveys are needed in portions or throughout the species' range. For the Allegheny woodrat, systematic field surveys have been conducted in Indiana, Maryland, New Jersey, New York, Ohio, Pennsylvania, and Virginia, and are part of the Wildlife Action Plans of some other states, but no comprehensive, integrated assessment has been conducted throughout the species' range (Wright Chapter 1, Peles and Wright Chapter 12).

(3) An assessment of the best available information, including literature and opinions of experts, to determine the most plausible causes for the species' decline. Inherent to this process is assessing if causalities for the decline have been remedied to an extent that warrants proceeding with the reintroduction project. Likely causes for extirpation or declines of some woodrat populations are beginning to emerge and form a basis for this type of assessment (LoGiudice 2006, Chapter 2).
(4) The identification of general areas within the species' historic range and geopolitical region of conservation jurisdiction (unless a federally endangered or threatened species, the maximum extent of this delineation typically will be defined by state boundaries) warranting consideration as a reintroduction site.
(5) A list of criteria to be applied for identifying preferred sources of individuals for use in the reintroduction, including an assessment of the conservation status, and a review of the genetic and phenotypic characteristics of individuals comprising a prospective source population.
(6) The identification of collaborators and stakeholders with appropriate levels of expertise, interest, and resources for contributing to the reintroduction project.
(7) A list of conservation authorities and their respective jurisdictions legally responsible for authorizing and supporting the various aspects of a reintroduction project. Essential to this process is being aware that permits (and, typically, other forms of formal approval) are required to legally capture and handle wildlife (ASM Animal Care and Use Committee 1998).
(8) An itemized budget of anticipated costs, and an associated review of potential sources of funding and other support.
(9) A general timeline for implementing and completing various phases of the project.

Implementation

Implementation of a reintroduction project follows satisfactory completion and interpretation of preliminary assessments, and presumes that the outcome of the feasibility study warrants initiating a project. The implementation of a well-conceived reintroduction project should comprise pre-translocation, translocation, and post-translocation phases. Although often portrayed as being discretely situated within the project, the various phases inherently are interdependent and often overlap considerably during the reintroduction process.

Pre-translocation Phase

The primary components of the pre-translocation phase of a reintroduction project consist of: (1) selection of reintroduction sites; (2) selection of sources of animals for translocation; and (3) preparation and implementation of an outreach program (human dimensions). During this phase of the project guidelines established in the feasibility study are applied and further refined for the process of selecting reintroduction sites, identifying sources of animals, and establishing interactions with collaborators, stakeholders, and the public at reintroduction areas.

Reintroduction Site Selection

A suitable release site must be within the historic range and possess habitat qualities presumed important for sustaining a population of the species being reintroduced. Although specific habitat components differ among the various species of woodrats, generally most are associated with readily discernable macro-features (e.g., talus slopes on ridges; Balcom and Yahner 1996, Ford et al. 2006) distributed as isolated units within the landscape. The advent and refinement of geographic information system (GIS) technology has considerably enhanced the capabilities for efficiently assessing and displaying the dispersion and connectivity of macro-habitat components within the landscape (Kiester et al. 1996). These capabilities are particularly appealing for application in selecting reintroduction sites for species using habitats in a coarse-grained manner (Wiens 1976), such as Allegheny woodrats, which necessitates detection and assessment of relatively rare habitat components discretely structured within the landscape. The process of selecting reintroduction sites should proceed in a hierarchical fashion, from coarse- through fine-level GIS assessments to onsite visits of prospective sites.

Fundamental to coarse-level assessment is the acquisition of GIS layers (i.e., electronic files containing specific features in the landscape, such as habitat types, geopolitical boundaries, and specific points of interest) comprising locations of active and historic woodrat sites and macro-habitat features identified as important in the area being evaluated for the reintroduction. For example, important macro-habitat features identified as important to Allegheny woodrats in the mid-Atlantic region that could be represented in GIS layers include forests, rocky-outcroppings, and the distribution of public lands (Ford et al. 2006). Overlaying these features (portraying the various GIS layers simultaneously on a single projection) with the locations of active and historic population sites enables a concise depiction of the juxtaposition of important habitat components in relation to areas of known occupancy (current and past) by woodrats. A coarse-level assessment provides a basis for excluding large portions of the landscape from further consideration as potential reintroduction sites and for identifying areas with macro-habitat features worthy of finer-level consideration as prospective reintroduction sites. The development of GIS-based landscape models has been accomplished in several states for woodrat species, including *N. magister*, as part of the nationwide GAP analysis project (Scott et al. 1993, National Biological Information Infrastructure 2007). These models were based on a small suite of general habitat variables and, thus, lack specificity in identifying portions of the landscape suitable for woodrats. Nonetheless, the GAP-derived models provide a basis for guiding the development of GIS models more specific to woodrats, which has been the case in Pennsylvania (Mengak et al. Chapter 7). A spatially explicit GIS-based habitat model developed for Allegheny woodrats in the Daniel Boone National Forest, Kentucky, incorporated a more detailed set of landscape-level variables that proved useful in predicting the location of active woodrat sites and serves as an excellent example of how this technology could be applied for remotely identifying prospective reintroduction sites (Ivanovich 2000).

Many of the general habitat features assessed at the macro-habitat scale can also be interpreted and used in fine-level assessments. As an example, a forest layer of a GIS often comprises sublayers such as forest type (i.e., deciduous, conifer, or mixed) and forest age class. Similarly, public lands usually can be further categorized by management activities (e.g., natural areas where no timber harvest is permitted or areas designated for multiple-use, where the sustainable harvest of timber is permitted). The availability of sublayers for the various general attributes provides an opportunity to more thoroughly identify specific portions of the landscape that are likely to be suitable as reintroduction sites. The GIS analysis should progress to the identification of specific sites possessing a suite of attributes likely to support a population of woodrats, and then an evaluation of the manner in which sites with these attributes are dispersed (e.g., degree of clustering or isolation) within the prescribed assessment area. The outcome of the GIS analysis should include the following:

(1) Point locations for all sites possessing macro-habitat features known to be associated with woodrat population sites (hereafter these point locations are referred to as prospective sites).
(2) An evaluation of specific (finer-level) attributes associated with prospective sites. Generally, however, the attributes being assessed will be related to the availability of food and cover. For example, mast producing tree species such as white oak (*Quercus alba*), are a reliable source of food (Castleberry et al. 2002a), and rocky outcroppings at ridgetops and along gorges provide important cover (Balcom and Yahner 1996, Castleberry et al. 2002c, Ford et al. 2006) for woodrats in the mid-Atlantic region and should be considered as positive attributes for a prospective reintroduction site.
(3) An interpretation of habitat conditions surrounding prospective sites. Of particular importance in this assessment is the distance that forested conditions to extend beyond a prospective site and the degree to which the surrounding forest is fragmented (usually assessed within various buffered radii distances from the prospective site). The implication for this assessment is based on evidence that edge-tolerant predators will occur at higher levels in fragmented landscapes than in large, contiguous forest tracts, and that their presence is presumed to have deleterious consequences for woodrats (e.g., direct predation by great horned owls (*Bubo virginianus*) or transmission of the roundworm (*Baylisascaris procyonis*) through contact with raccoons (*Procyon lotor*) (McGowan 1993, Birch et al. 1994), as well as evidence that in some parts of the range, Allegheny woodrat population persistence is associated with a broad buffer of intact forest (Hassinger et al. Chapter 8)).
(4) An assessment of conditions favorable for supporting dispersal of woodrats and the likelihood that dispersing individuals will encounter areas with habitat

conditions suitable to sustain a local population. This assessment also pertains to the discrete placement of multiple reintroduction sites in an area with the intent of establishing or maintaining a metapopulation (a population segregated into semi-isolated units connected by corridors able to facilitate some level of exchange among individuals; Opdam 1991, Wells and Richmond 1995, and Litvaitis and Villafuete 1996). An alternative but similar strategy would be the purposeful selection of a reintroduction site with the intent of enabling the eventual interaction of individuals with those from existing populations. In this case, an important goal would be to select reintroduction sites linked to other reintroduction sites or established population sites by habitat conditions appropriate to function as movement corridors. Wright (1998) identified research topics that should be addressed to enable conservationists to more effectively incorporate metapopulation concepts into strategies for reintroducing Allegheny woodrats. These strategies emphasize the importance of better understanding dispersal capabilities by age class and gender, habitat conditions contributing to successful dispersal, the extent that sites within a metapopulation are linked by dispersing individuals, and natural patterns of extirpations and recolonizations within a metapopulation.

(5) Determination of the conservation status of lands (public or private) at a prospective site, within radii of various predetermined distances buffering a prospective site, and within the boundaries of potential corridors connecting to other prospective sites or established sites. Generally, public lands or lands owned by conservation organizations should be preferred as reintroduction sites because of the prospect for maintaining the long-term integrity of habitats in these areas (i.e., development is less likely than on private lands) and greater opportunity to manage portions of the landscape for the conservation of woodrats.

Unfortunately, specific guidelines have not been developed to guide the management of habitats for woodrats at either fine or coarse scales. Existing habitat studies provide a basis for guiding the process of identifying and protecting habitats important for Allegheny woodrats (Balcom and Yahner 1996, Ivanovich 2000, Castleberry et al. 2002c, Ford et al. 2006, Castleberry Chapter 4). Hassinger et al. (Chapter 8) have proposed a habitat management plan for application in the general conservation or reintroduction of Allegheny woodrats. The plan suggests that habitat protected for Allegheny woodrats should comprise an appropriate rock habitat surrounded by a 200-m-wide foraging zone, where management actions are designed to improve production of hard mast, and then by an outer 2-km-wide landscape protection zone or nondissection buffer where commercial forest management may occur but roads, right-of-ways, and other forms of permanent fragmentation and dissection are discouraged. However, the authors acknowledge that these criteria are based on interpretation of existing habitat studies and not an empirical assessment of outcomes based on the application of guidelines that have actually been applied. Thus, the plan should be regarded as an initial, general guideline, which can be adaptively modified based on assessments of the manner in which native or reintroduced woodrat populations respond to this type of habitat management.

Assessment of the GIS analysis should yield an objective ranking of the quality of prospective sites based on biological and ecological criteria identified in the feasibility study. Ultimately, the selected sites should possess qualities that meet biological criteria for sustaining a viable woodrat population at the release location, provide conditions that facilitate opportunities for woodrats to disperse and pioneer new areas (i.e., establish a metapopulations structure or, similarly interact with other reintroduced populations or established populations in the region), and are likely to retain conditions appropriate for maintaining woodrat populations (metapopulations) in the future (i.e., changes in patterns of land use that would be detrimental to the population are unlikely based on current land-use designations).

The final selection of a reintroduction site must be determined through onsite inspections of prospective sites ranked most favorably by GIS analysis. The inspectors should include experts that have developed, through training and experience, a conceptual model of conditions appropriate for sustaining woodrats. The value of these opinions should not be underestimated and should be given serious consideration in the final determination of a reintroduction site.

Source of Animals

Various criteria have been established for selecting sources of individuals for reintroduction projects. Ideally, reintroduced animals should be obtained from areas that represent the same subspecies that occurred at the reintroduction site (IUCN 1998). To enhance the likelihood of a successful reintroduction, Meffe (1995) suggested that reintroduced animals should be obtained from nearby populations that occupy a range of environmental conditions similar to those that will be encountered at reintroduction sites. The recommendation is based on the premise that founders from local populations have evolved under selective pressures imposed by the range of environmental conditions in the region and, as such, should be best suited to adapt to conditions imposed at release sites.

In general, translocation of individuals from nearby populations should be considered the most practicable and defensible approach for obtaining animals for use in a reintroduction project. However, defining what constitutes an appropriate, nearby population often can be confounded by a variety of biological and practical issues. Inherent to this approach in defining a source of individuals for translocation is the assumption that populations geographically close to the reintroduction site are composed of individuals of similar genetic composition to that of the extirpated population. This assumption can be compromised somewhat by organisms, such as Allegheny woodrats, that exist as disjunct aggregations among suitable habitats dispersed throughout the landscape (Balcom and Yahner 1996, Ford et al. 2006). The natural isolation of populations through discrete use of the landscape and limited ability for juvenile woodrats to disperse through suboptimal habitats undoubtedly contributed to genetic uniqueness among populations—even those occupying the same general region. For example, analysis of genetic variability among Allegheny woodrats in West Virginia demonstrated considerable differentiation among subpopulations separated by as little as 3 km (Castleberry et al. 2002b).

Consequently, identifying historic patterns of gene flow is a challenge for interpreting the degree of genetic similarity that existed among extant and extirpated populations. The genetic issue is complicated further by changes (declines and isolation) in woodrat populations that have been brought about by anthropogenic alterations to the landscape (Balcom and Yahner 1996, Ford et al. 2006, Smyser and Rhodes Chapter 9). These types of changes in the characteristics of a population can contribute to substantial reduction in genetic variation through bottlenecks (rapid reduction in the size of a population followed by some level of recovery; Bonnell and Selander 1974, O'Brien et al. 1983), or through genetic drift (Gilpin and Soulé 1986), which is most pronounced in small, isolated populations. Therefore, selecting a nearby, apparently viable population as the primary source of animals for the reintroduction (regardless of the current size of the population) may be contributing only a portion of the original levels of genetic diversity contained within woodrat populations from the region. The situation is further confounded by the likelihood that only a portion of the genetic diversity contained within the source population will be sampled among translocated individuals (best thought of as a type of sampling error). Following this scenario, the reintroduced population may have considerably lower levels of genetic diversity than would be considered desirable for the long-term sustainability of the population. Smyser and Rhodes (Chapter 9) discuss in detail the complexities of choosing a genetically appropriate source population for a proposed reintroduction of Allegheny woodrats in Indiana.

Basic to any reintroduction project is the identification of source populations capable of supporting the sustained removal of sufficient numbers of individuals to meet biological and practical components of the project. This criterion has implications related directly to the conservation of source populations (i.e., removal of individuals will not adversely impact the long-term sustainability of genetic integrity of the population) and for realistically meeting projected timelines for completing the translocation phase of the project (i.e., a sufficient number on individuals will be available to sustain a meaningful reintroduction).

Coordinators of reintroduction projects should be aware that identifying the best source (or sources) of individuals for use in a reintroduction project often is subject to various interpretations—regardless of the quality of biological and conservation information available for the species to be reintroduced. Consequently, disagreements as to what constitutes an appropriate source population are not unusual and should be anticipated (Serfass 1998), especially given the complexity of integrating the various biological and management criteria into a cogent plan for conducting a reintroduction project.

Number of Animals

Reintroduction programs need to incorporate minimum viable population (MVP) criteria when considering the number of animals needed to be released to establish and maintain a population. MVP considerations include loss of genetic diversity, demographic stochasticity, and environmental stochasticity (Gilpin and Soulé 1986). Meeting MVP criteria in reintroduction projects is dependent on the number of animals released, demographics and post-release survival of reintroduced individuals, and genetic composition of the source populations. Unfortunately, identifying the number of individuals necessary to ensure the establishment of a viable, self-sustaining population is fraught with uncertainties related to the responses of translocated individuals to conditions at the reintroduction site (e.g., post-release survival rates and levels of reproduction).

The low survival rates characteristic of *r*-selected organisms adds further complications in determining the number of animals needed for release. As an example, the Allegheny woodrat in Pennsylvania has annual survival and mean life expectancies estimated at only 0.46 and 1.3 years, respectively (J. Hassinger unpublished data), and does not readily breed during the winter. Reintroduction efforts must take into account potential low over-winter survival while planning for the number of animals to be released.

There are few examples for small mammals (or specifically woodrats) to guide the determination of what constitutes an adequate number of woodrats to be released at a reintroduction site. Jordan (2003) suggested that release of at least 30–50 individuals (sometimes higher) is normal among the few reintroduction projects that have involved small mammals. Unfortunately, woodrats are territorial and tend to naturally maintain relatively small population units. For example, the typical population size for Allegheny woodrats is less than 20 individuals—even for healthy, self-sustaining populations (Mengak et al. Chapter 7). A population of this size is unlikely to be able to adequately buffer against stochastic fluctuation and catastrophes. Historically, populations of Allegheny woodrats likely have been subjected to episodic bouts of extirpation and subsequent recolonization by dispersing individuals from surrounding populations. Attempts to reintroduce the number of woodrats at a site within the natural size of a typical population might be predisposed to failure because of initial stochastic processes (e.g., differential survival by gender or loss of genetic diversity if a few individuals contribute disproportionately to breeding). Release of a larger group of individuals (e.g., >20 individuals) to overcome could contribute to social stress because of the innate territoriality of woodrats (Kinsey 1977, Peles and Wright Chapter 5).

These aspects of Allegheny woodrat biology confound decision-making regarding the appropriate number of individuals that should be released to enhance the likelihood of establishing a population. Considering that Allegheny woodrats have short life expectancies and naturally fluctuating small populations, including bouts of extirpations and natural recolonizations, a strategy of pulsed releases (supplementations) over time may be necessary to provide numeric and genetic enrichment to sustain a reintroduced population. Such a strategy could be designed to emulate interactions among woodrat populations typically facilitated by dispersing individuals. Although much remains to be learned about natural dispersal of Allegheny woodrats, apparently, adult woodrats of both sexes are potential dispersers (Wood Chapter 3). Supplementations would be continued until dispersing individuals from the reintroduction site pioneered suitable habitats in the region, thus contributing to the establishment of interacting populations typical of self-sustaining woodrat populations. Alternatively, a similar outcome might be achieved by releasing individuals among several clustered reintroduction sites in a manner intended to enable eventual dispersal among populations. At the least, protocols should be in place to supplement reintroduced populations if evidence from monitoring indicates a need because of unexpectedly high levels of post-mortality or the dispersal of a large number of individuals to locations distant enough whereby they are functionally no longer part of the reintroduced population.

Application of population viability analysis (PVA) has considerable value for establishing a basis of what constitutes an adequate number and demographic composition of individuals to be released, and for assessing temporal and spatial strategies for releases (White 2000). PVA can help ensure, with reasonable certainty, that the reintroduced population will be large enough to avoid adverse consequences from short-term demographic stochasticity, long-term genetic stochasticity, and will develop into a sustainable MVP (Shafer 1981). McCleery et al. (2005) provided an excellent example of the application of PVA simulations for anticipating outcomes of translocation projects—in this case, for determining the likelihood that the Key Largo woodrat (*N. floridana smalli*) population would derive long-term benefit by supplemental stockings. The simulations indicated a high probability that the Key Largo woodrat population would become extinct unless management activities were implemented to mitigate deleterious impacts of limiting factors—habitat alterations, black rats (*Rattus rattus*), fire ants, and feral cats have been suggested plausible reasons for declines in the population. However, the trajectory of the population decline was reduced when supplemental releases were included in the model, and simulations that included a scenario of ongoing supplementations substantially reduced the likelihood of extinction.

Releasing enough individuals to ensure adequate levels of genetic diversity within a released population should be an important goal of a reintroduction project. Leberg (1990) applied population genetic principles developed by Wright (1969) to demonstrate that loss of heterozygosity in founding populations is small if the population consists of 10 or more individuals of equal sex ratios. However, the adequacy of reintroducing this few individuals presumes high survival among founding individuals, and that all or most contribute to breeding and production of offspring. These presumptions are likely to be unrealistic, and the release of this few animals may be insufficient to establish a self-sustaining population, especially for species having high mortality rates (*r*-selected species), such as woodrats.

Designing release strategies to facilitate the establishment of a metapopulation structure through the release of animals at two or more spatially disjunct areas connected by corridors is likely advantageous for the long-term persistence of the reintroduced population and could serve as an alternative or be integrated with supplementations. The strategy has various advantages, including the recreation of more natural processes whereby numeric and genetic enrichments are facilitated by dispersal of individuals among units of the metapopulation.

Human Dimensions

Another component that needs to be considered at the start of a reintroduction project is the development of a well-conceived public relations program. Project coordinators must recognize that a successful reintroduction requires more than making appropriate decisions regarding the biological and ecologically requirements of the species. Engendering support and interest for the species with the public (particularly, but not exclusively at the reintroduction area) is important for enhancing the likelihood of positive interactions and support during all phases of the reintroduction project. Coordinators of successful reintroduction projects also embrace opportunities for collaboration with professionals responsible for conservation activities at or near the reintroduction site.

Public Relations

Outreach, with the goal of building public support, arguably represents the most viable approach for achieving positive, long-term outcomes for important conservation issues, including the re-establishment of species by reintroduction (Kleiman 1989, Kleiman et al. 1996). Unfortunately, conservation professionals too often neglect or are unaware of the positive benefits that can be derived from such efforts.

The value of public education has been demonstrated in various reintroduction projects, most notably those involving larger species of wildlife, such as wolves (Fritts et al. 1997) and river otters (Serfass 1998, Serfass et al. 2003), that are considered appealing to the general public. Admittedly, the public typically is less familiar and generally uninformed about woodrats and other smaller animals. However, the Allegheny woodrat possesses many characteristics that can be used to garner interest in the species (including unique behaviors, such as the creation of middens), and thus is a good candidate on which to focus public education events (e.g., slide shows about its natural history and conservation status to school groups or conservation groups). Also, the inherent message associated with a reintroduction project—a species has become imperiled and an effort is underway to restore it—is likely to attract the public's attention, and thereby provide another opportunity to convey an important conservation message. For the woodrat, this message should include discussions about human-induced changes in the landscape that have contributed to its decline. Ultimately, the quality of the message will depend upon the creativity of the conservationists conducting the reintroduction project.

Development of the outreach program should coincide and integrate with the initial planning of the reintroduction project and include the following steps:

(1) *Identify the audience*—The intended audience for reintroduction projects is generally the population living around the release site. However, conservation organizations, groups, civic leaders, and school conservation clubs (among others) should be informed about the project and included in educational events. The process should also seek to identify any individuals or groups that may have concerns (or express concerns) about the reintroduction project.
(2) *Identify the purpose and specific messages to be portrayed*—The initial message typically will be about various aspects of the project (justifications, objectives, and goals) and the biology and natural history of the animal to be reintroduced. The message must emphasize that Allegheny woodrats, unlike Norway rats

(*Rattus norvegicus*), are a native species and will not become a pest species. Specific messages should be developed to coincide with different phases of the reintroduction project. At times, a message may need to be developed in an adaptive manner to communicate unexpected events or to address a particular concern or question posed by the public.

(3) *Determine the best mechanisms available for communicating the message*— Typically, an initial message can be delivered to communities near reintroduction areas through local media (i.e., newspapers, radio, and television). Preparation of press releases and media packet is useful in guiding interviews with reporters and ensuring that important facts about the project are appropriately portrayed. Activities involving capture, handling, and release of animals provide unique opportunities for participating media to obtain images of animals being reintroduced and a more intimate understanding of the reintroduction process. The public relations program should be ongoing and designed to inform the public about progress and important events associated with the various phases of the project.

Professional Interactions

Reintroductions are multi-faceted efforts requiring various levels of interaction among a variety of conservation professionals. Procedures for disseminating information among members of the reintroduction team and to other professionals associated with the project should be established early in the program (prior to initiating trapping and translocations). Depending on the conservation status of the species, state and/or federal permits will be required for capture, handling, and translocation. In many cases, reintroductions take place on public lands or lands owned and managed by private conservation organizations (see Site Selection). Consequently, members of the reintroduction team should anticipate frequent interactions with professionals responsible for management of the lands at the reintroduction site. In some cases there may be formal expectations regarding the reporting of activities associated with the reintroduction to agency representatives responsible for managing the lands. Regardless, professional courtesy warrants effective and frequent communication with others directly or indirectly associated with the project.

Translocation Phase

Capture and Handling

High survival rates among founding individuals (with respect to what would be anticipated for the species being reintroduced) are essential for enhancing the likelihood of a successful reintroduction program. The prospects for high levels of survival during stressful, post-release acclimation periods likely will be enhanced when reintroduced animals are in good physical condition, which is influenced by protocols implemented during capture and handling (e.g., removing from trap, tagging, transport, and captive care).

Animals are subjected to stress and potential injury when captured and handled with clinical consequences manifested as a complex sequence of conditions, including external trauma (e.g., trap-related injury) and skeletal muscle necrosis (Serfass et al. 1993, Hartup et al. 1999). The degree of stress and frequency of injuries can be mitigated by following the best available guidelines for the capture, handling, and care of the species (i.e., from review of literature and consultation with experts; ASM 1998). Regardless, responses of animals to capture and handling can be unpredictable and some level of stress or the occurrence of injuries should be anticipated. Mechanisms should be in place to assess an animal's condition through the various phases of the translocation process. Ideally, animals should receive an immediate, post-capture physical exam to identify injuries in need of veterinary care, and to serve as an assessment of capture and handling procedures.

Following completion of physical exams, animals should be transported to captive facilities for a period of confinement to facilitate veterinary evaluation and, if necessary, receive treatment for injuries or disease prior to being released. Protocols for transport and captive management will vary by species and should be developed in collaboration with veterinarians participating with the reintroduction project. The type (e.g., size of caging and level of quarantine) and duration of confinement initially will be defined by goals of the project (e.g., if the intention is to simultaneously release a large group of animals, some individuals may require a long period of captivity until the desired number of animals is attained), and initial assessments of captive requirements for the species and responses of individuals to captive environment (e.g., some species may respond poorly to extensive periods in captivity, whereas others may thrive in captivity), and subsequently refined through ongoing assessments and interpretations of individual responses to captive conditions. Allegheny woodrats generally appear to adapt readily to captive conditions and do not require elaborate housing (Poole 1940, Kinsey 1976). However, adults must be confined individually to prevent aggressive interactions, which can result in severe injuries or deaths.

The benefits and justifications for incorporating a captive management program generally are well recognized and typically recommended to be incorporated into the design of reintroduction projects (Hoover et al. 1984, Kleiman 1989, Dodd and Seigel 1991, IUCN 1998). The primary objectives of confining animals prior to reintroduction should be to identify and treat injuries, diseases, and other forms of general pathology, and to enhance the general condition of animals by feeding a nutritionally complete diet—the ultimate intention being to promote the release of healthy, vigorous animals that have a high probability of survival (Serfass et al. 1993). The potential for spreading infectious disease and parasites is a particularly important issue for the translocation of wild animals, further justifying the implementation of a captive management program to facilitate health evaluations (Viggers et al. 1993, Serfass et al. 1995). In the western United States, woodrats occasionally harbor Lyme disease and hantavirus pathogens (Lane and Brown 1991, Dearing et al. 1998). Allegheny woodrats could be susceptible to these diseases, but there are no documented reports. The susceptibility of Allegheny woodrats to raccoon roundworm is, however, well established (LoGiudice 2000). Although direct transmission of *B. procyonis* from woodrat to woodrat does not appear feasible, translocation of an infected woodrat and its subsequent ingestion by a raccoon could spread this parasite to new areas (LoGiudice 2000). Finally, holding animals in captivity provides opportunities to schedule releases to coincide with favorable environmental conditions.

Considerations for Capture and Restraint

Various box and cage-type traps have been used successfully to capture woodrats. The use of Sherman traps (model PCLF15; H. B. Sherman Traps, Inc., Tallahassee) and Tomahawk live traps (model TL201; Tomahawk Live Trap, Tomahawk, Wisconsin, USA) has been reported for woodrats (McCleery et al. 2006, Parker 2006, Mengak et al. Chapter 7). For restraint during handling, Frase and Van Vuren (1989) and Parker (2006) assessed the use of ketamine hydrochloride and isoflurane, respectively. Unfortunately, there have been no published studies comparing the efficacy of various traps, restraints, transport techniques, or captive management procedures for woodrats. However, numerous general models exist for the handling and care of small mammals (Kleiman et al. 1996), which can serve as initial guides in developing or enhancing procedures for woodrats.

Ideally, trapping should be avoided during primary periods of primary reproductive activity (spring) and of low food production (i.e., late fall through early spring). First litters of Allegheny woodrats typically are born in mid-spring, and most live-trapping programs are designed to avoid the peak in lactation during this period (Mengak et al. Chapter 7). However, breeding has been documented throughout summer and into fall (Castleberry et al. 2006) and, consequently, trapping during some periods of the reproductive season is unavoidable if woodrats are to be obtained for release when natural foods are most available at reintroduction sites.

Release Strategies

Plans to reintroduce animals should consider factors such as the time of year (implications related to the availability of food and opportunities to prepare a winter food cache) and behavioral characteristics of the species. Seasonal variation in the availability of food will be more pronounced for species occupying northern portions of the woodrat's distribution. Woodrats begin assembling a winter food cache in mid-summer and typically do not change dens after early fall (Peles and Wright Chapter 5), so the chance to accumulate a sufficient cache to survive the winter is likely to diminish as release periods extend into later summer and fall. The optimum period for capture and release may thus be mid-summer, unless released individuals are provided with over-winter food supplements.

Survival of Allegheny woodrats reintroduced in Ohio was higher in summer (after leaf-out) than during spring (prior to leaf-out), presumably because of increased vegetative cover, which provided better avoidance from predators (Schlie

1985). However, interpretation of this outcome is confounded because of different release strategies used for the two seasons—"direct" releases (animals released directly at the release site 2–3 days post-capture) were used during spring and "gentle" releases (animals held in cages at release site for ≥7 days post-capture and supplemental food was provided following release) in summer. McGowan (1993) demonstrated that territorial behavior of woodrats should be considered in release strategies. In this case, a phased-release strategy (animals not simultaneously released at the site) appeared to contribute to greater dispersal (and associated higher mortality) than that observed when individuals were released simultaneously. Presumably, the greater movement and lower survival among individuals in the phased-release group was related to agonistic encounters between established woodrats (those released earliest in the project typically established territories near the release site) and individuals released later in the project, which typically dispersed furthest from the release site. However, further study is needed before concluding that release strategy was the primary casual mechanism for the observed movement patterns.

An important consideration is the implementation of hard-release (i.e., animals directly transported from the capture site and released) or soft-release (i.e., animals transported from capture site to a reintroduction site where they are housed in cages for a period of adjustment until being released) strategies. The higher survival rate among woodrats released in Ohio during summer in comparison to those released during spring (Schlie 1985), which was attributed to seasonal differences in vegetative cover, could likely have been from the spring use of hard releases in comparison to summer soft releases. The simultaneous release of all (most) individuals comprising the reintroduced population has been suggested as the most appropriate strategy for releasing woodrats at a particular site (McGowan 1993), which is most easily accomplished by a soft-release. However, exclusive use of a simultaneous release strategy would prohibit the use of strategies that incorporate supplemental releases. Failure of previous efforts to successfully reintroduce Allegheny woodrats suggests that alternative strategies, including supplemental releases, may be warranted in future reintroduction projects. Clearly, additional investigation is required to determine what scenarios constitute the best strategies for releasing woodrats, including evaluations of approaches that could be used to minimize agonistic encounters between established woodrats and new individuals released to supplement the population.

Post-translocation Phase

Evaluation and Monitoring

Monitoring initial fates of translocated wildlife and subsequent long-term studies to determine if self-sustaining populations become established should be an important aspect of reintroduction projects (IUCN 1998, Serfass et al. 2003). The level of funding will dictate the type and intensity of effort that can be devoted to post-release monitoring. Regardless, monitoring should incorporate an initial, post-release (short-term) and ensuing long-term phase.

Short-Term Phase

The primary goals of short-term monitoring are to determine survival rates, movement patterns, and levels of interaction among translocated individuals. Minimally, some mechanism should be applied that enables identification of each reintroduced individual. Numerous techniques have been devised that are applicable for marking small mammals, including ear-tagging, toe-clipping, implantable (subcutaneous) transponder chips, and developing a genetic profile for each individual from tissues collected during capture and handling (Paetkau 2004, Peakall et al. 2006). Typically, metal ear tags or ear tattoos have been used to identify individual woodrats (Mengak et al. Chapter 7).

Although expensive in time, labor, and equipment, radiotelemetry is the most effective method for determining the fates of reintroduced individuals and has been used in all Allegheny woodrat reintroduction efforts. Success of the reintroduction project depends on a sufficient number of interacting individuals persisting at the reintroduction location to achieve reproductive success. Thus, information about post-release survival and movements is essential for informed decision making about the short-term prospects for a successful reintroduction and determining if there is a need to adjust the reintroduction process. Radiotelemetry is the only mechanism for reliably assessing important demographic and behavioral characteristics of a species, such as survival, reproduction, movement patterns, interactions with conspecifics, and determining causes of deaths. Also, if supplemental releases occur during the project, radiotelemetry would be an ideal method for assessing if survival and movement of newly released animals is influenced by agonistic encounters with resident animals.

Long-Term Phase

The general intent of long-term monitoring will be to determine if the reintroduction results in the establishment of a self-sustaining population. Therefore, long-term monitoring ideally should continue until a definitive status is determined for the population (e.g., the area is judged unsuitable for sustaining a viable population or, alternatively, the population is considered secure). A properly designed live-trapping effort conducted at five-year intervals would likely yield sufficient information to determine the general status of the population (e.g., evidence of persistence and demographics), but would be unlikely to provide information adequate for a meaningful assessment of population trends or a valid interpretation of factors contributing to those trends (e.g., unexpectedly high fecundity or, conversely, mortality). Also, researchers involved in long-term assessments must be aware that multi-year studies of natural Allegheny woodrat populations demonstrate considerable levels of stochastic fluctuation in population size among years, including extirpation and subsequent recolonization (Mengak et al. Chapter 7). Such naturally occurring population fluctuations can confound the interpretation of results derived from monitoring reintroduced populations. Consequently, monitoring must take place for a period of time sufficient for researchers to determine if changes in the population size (increasing or decreasing) are the result of natural cycling or an indication that the reintroduction is actually succeeding (population stable or expanding) or failing (population declining).

Often overlooked in the monitoring process are opportunities for long-term assessments of the population's genetic composition. An apparently thriving reintroduced population (based on population size) may, nonetheless, have lost a substantial portion of genetic diversity, which could have unforeseen deleterious effects to the population. Elk in Pennsylvania is an excellent example of a reintroduced population that has increased in numbers, but retained very little of the genetic diversity retained in the source population (Williams et al. 2002). Useful information could be obtained by collecting samples for genetic analysis at any stage of the monitoring process. However, a preferred scenario would be to obtain samples from all individuals comprising the reintroduced population and then sequentially during various phases of the monitoring process. This approach would enable an assessment of retention or loss of genetic variability over time and, thus, enable informed decision making regarding possible conservation actions such as supplementations.

Conservation and Management of Declining Species

The ranges of some species of woodrats have suffered considerable declines, presumably because of the direct and indirect consequences of anthropogenic disturbances. These declines have prompted discussions about implementing more active management approaches for woodrats, including implementing reintroduction projects. Species of small mammals exhibiting *r*-selected traits have received very little active conservation attention in comparison to larger, charismatic species, and there are few examples of them being the focus of reintroduction projects for woodrats (Schlie and Bookhout 1985, McGowan 1993, Gerber et al. 2003, McCleery et al. 2005) or other *r*-selected small mammals (Jordan 2003). Consequently, there are few specific reintroduction models available to guide the process of reintroducing of woodrats or for anticipating the most likely outcomes of reintroduction projects. Nonetheless, past failed attempts to reintroduce woodrats provide useful insight into the reintroduction process. For example, Schlie (1985) advocated release of woodrats after leaf-out and the use of a soft-release strategy; McGowan (1993) and LoGiudice (2003) demonstrated the importance of avoiding *B. procyonis* contamination at the release site; and Wright (1998) discussed possible social interactions that may discourage individuals from settling at a site when a pulsed release is implemented. Although these post-hoc assessments are valuable, a comprehensive approach must be developed to enhance the likelihood of successfully reintroducing woodrats. Thus, an important challenge for woodrat conservation will be the development of a coherent reintroduction strategy derived from the ongoing development, implementation, assessment, and subsequent refinement of reintroduction concepts and techniques.

The IUCN-RSG (1998) has developed useful general guidelines for reintroduction projects. However, specific components of the reintroduction process must be designed to conform to the unique requirements of the species being reintroduced. For woodrats (and the case for most *r*-selected species) there have been no comprehensive reintroduction efforts from which the reintroduction process can be modeled. In this chapter, I put forth a complete conceptual model to serve as a basis for developing and refining approaches that can be applied and evaluated for the reintroduction of woodrats. Although similar in many aspects, the recommendations diverge from approaches typically applied for reintroducing larger, *K*-selected mammals (which represent the majoring of mammalian reintroductions) by emphasizing challenges related to reintroducing the Allegheny woodrat—an organism that occupies habitats typically dispersed as small patches within the landscape, maintains relatively low population densities among interconnected populations (metapopulations), but otherwise has *r*-selected characteristics, including high rates of mortality. Specifically, I recommend an approach that recognizes the obstacles of reintroducing woodrats, whose existence revolves around cyclic episodes of extirpations and recolonization from individuals dispersing from adjacent populations, as would be expected for an organism that has evolved to persist through interactions within a metapopulation. Inherent to this consideration is attempting to recreate or maintain a metapopulation as part of the reintroduction strategy. Such an approach is likely to comprise multiple scenarios, including, but not limited to: (1) reintroducing a new population in the proximity of extant populations with the intent of fortifying an existing but deteriorating metapopulation; (2) reintroducing several populations in proximity to one another with the intent of establishing a metapopulation; or (3) reintroducing a single population with ongoing supplementation of new individuals until dispersing individuals colonize new areas, resulting in the establishment of a metapopulation. Regardless of the scenario, high mortality should be anticipated and ongoing supplementations should be considered for the design of future reintroduction projects involving woodrats.

Conservationists have much to learn about the best procedures to apply for reintroducing Allegheny woodrats. Designing future woodrat reintroduction projects as series of well conceived, noninvasive experiments will facilitate objective evaluation and interpretation of the various phases of a restoration effort. Such an approach will serve as a basis for refining the reintroduction process for woodrats, as well as other similarly adapted *r*-selected species.

References

ASM Animal Care and Use Committee. 1998. Guidelines for the capture, handling and care of mammals as approved by the American Society of Mammalogists. Journal of Mammalogy 79:1416–1431.

Balcom, B. J., and Yahner, R. H. 1996. Microhabitat and landscape characteristics associated with the threatened Allegheny woodrat. Conservation Biology 10:515–525.

Birch, G. L., Feldhamer, G. A., and Dyer, W. G. 1994. Helminths of the gastrointestinal tract of raccoons in southern Illinois with management implications of *Baylisascaris procyonis* occurrence. Transactions of the Illinois State Academy of Science 87: 165–170.

Bonnell, M. L., and Selander, R. K. 1974. Elephant seals: genetic variation and near extinction. Science 184:908–909.

Castleberry, N. L., Castleberry, S. B., Ford, W. M., and Wood, P. B. 2002a. Allegheny woodrat (*Neotoma magister*) food habits in the central Appalachians. American Midland Naturalist 147:80–92.

Castleberry, S. B., King, T. L., Wood, P. B., and Ford, W. M. 2002b. Microsatellite DNA analysis of population structure in Allegheny woodrats (*Neotoma magister*). Journal of Mammalogy 83:1058–1070.

Castleberry, S. B., Wood, P. B., Ford, W. M., Castleberry, N. L., and Mengak, M. T. 2002c. Summer microhabitat selection by foraging Allegheny woodrats (*Neotoma magister*) in a managed forest. American Midland Naturalist 147:93–101.

Castleberry, S. B., Mengak, M. T., and Ford, W. M. 2006. *Neotoma magister*. Mammalian Species 789:1–5.

Dearing, M. D., Mangione, A. M., Karasov, W. H., Morzunov, S., Ottesen, E., and St. Jeor, S. 1998. Prevalence of hantavirus in four species of *Neotoma* from Arizona and Utah. Journal of Mammalogy 29:1254–1259.

Dodd, C. K., and Seigel, R. A. 1991. Relocation, repatriation, and translocation of amphibians and reptiles: are they conservation strategies that work? Herpetologica 47:336–350.

Ford, W. M., Castleberry, S. B., Mengak, M. T., Rodrigue, J. L., Feller, D. J., and Russell, K. R. 2006. Persistence of Allegheny woodrats *Neotoma magister* across the mid-Atlantic Appalachian Highlands landscape, USA. Ecography 29:745–754.

Frase, B. A., and Van Vuren, D. 1989. Techniques for immobilizing and bleeding marmots and woodrats. Journal of Wildlife Diseases 25:444–445.

Fritts, S. H., Bangs, E. E., Fontaine, J. A., Johnson, M. R., Phillips, M. K., Koch, E. D., and Gunson, J. R. 1997. Planning and implementing a reintroduction of wolves in Yellowstone National Park and central Idaho. Restoration Ecology 5:7–27.

Gerber, L. R., Seabloom, E. W., Burton, R. S., and Reichman, O. J. 2003. Translocation of an imperiled woodrat population: integrating spatial and habitat patterns. Animal Conservation 6:309–316.

Gilpin, M. E., and Soulé, M. E. 1986. Minimum viable populations: the processes of species extinctions. *In* Conservation Biology: The Science of Scarcity and Diversity, ed. M. E. Soulé, pp. 19–34. Sinauer Associates, Sunderland, Massachusetts.

Griffin, B., Scott, J. M., Carpenter, J. W., and Reed, C. 1989. Translocation as a species conservation tool: status and strategy. Science 245:477–480.

Hartup, B. K., Kollias, G. V., Jacobsen, M. C., Valentine, B. A., and Kimber, K. R. 1999. Exertional myopathy in translocated river otters in New York. Journal of Wildlife Diseases 35:542–547.

Hoover, J. P., Root, C. R., and Zimmer, M. A. 1984. Clinical evaluation of American river otters in a reintroduction study. Journal of the Veterinary Medical Association 187:1321–1326.

IUCN. 1998. Guidelines for Re-introductions. Prepared by the IUCN/SSC Re-introduction Specialist Group. IUCN, Gland, Switzerland and Cambridge, United Kingdom.

Ivanovich, E. S. 2000. The transferability of a predictive Geographic Information System model of Allegheny woodrat habitat in Kentucky. M.S. Thesis, Eastern Kentucky University, Richmond, Kentucky.

Jordan, M. J. R. 2003. Re-introduction protocols and considerations for small mammals. Re-introduction News 22:39–43.

Kiester, A. R., Scott, J. M., Csuti, B., Noss, R. F., Butterfield, B., Sahr, S., and White, D. 1996. Conservation prioritization using GAP data. Conservation Biology 10:1332–1342.

Kinsey, K. P. 1976. Social behaviour in confined populations of the Allegheny woodrat, *Neotoma floridana magister*. Animal Behaviour 24:181–187.

Kinsey, K. P. 1977. Agonistic behavior and social organization in a reproductive population of Allegheny woodrats, *Neotoma floridana magister*. Journal of Mammalogy 58:417–419.

Kleiman, D. G. 1989. Reintroduction of captive mammals for conservation. BioScience 39: 152–161.

Kleiman, D. G., Allen, M. E., Thompson, K. V., and Lumpkin, S. (eds). 1996. Wild Mammals in Captivity. The University of Chicago Press, Ltd. London, United Kingdom.

Lane, R. S., and Brown, R. N. 1991. Woodrats and kangaroo rats: potential reservoirs of the Lyme disease spirochete in California. Journal of Medical Entomology 28:299–302.

Leberg, P. L. 1990. Genetic considerations in the design of introduction programs. Transactions of the North American Wildlife and Natural Resources Conference 55:609–619.

Litvaitis, J. A., and Villafuete, R. 1996. Factors affecting the persistence of New England cottontail metapopulations: the role of habitat management. Wildlife Society Bulletin 24:686–693.

LoGiudice, K. 2000. *Baylisascaris procyonis* and the decline of the Allegheny woodrat (*Neotoma magister*). Ph.D. Dissertation, Rutgers, The State University of New Jersey, New Brunswick, New Jersey.

LoGiudice, K. 2003. Trophically transmitted parasites and the conservation of small populations: raccoon roundworm and the imperiled Allegheny woodrat. Conservation Biology 17:258–266.

LoGiudice, K. 2006. Toward a synthetic view of extinction: a history lesson from a North American rodent. BioScience 56:687–691

McCleery, R. A., Lopez, R. R., and Silvy, N. J. 2006. Population status and habitat selection of the endangered Key Largo woodrat. American Midland Naturalist 155:197–209.

McCleery, R. A., Lopez, R. R., Silvy, N. J., and Grant, W. E. 2005. Effectiveness of supplemental stockings for the endangered Key Largo woodrat. Biological Conservation 124:27–33.

McGowan, E. M. 1993. Experimental release and fate study of the Allegheny woodrat (*Neotoma magister*). Federal aid project report W-166-E-1, New York Department of Environmental Conservation, Delmar, New York.

Meffe, G. K. 1995. Genetic and ecological guidelines for species reintroduction programs. Journal of Great Lakes Research 21:3–9.

National Biological Information Infrastructure. 2007. The Gap Analysis Program. Available http://gapanalysis.nbii.gov.

O'Brien, S. J., Wildt, D. E., Goldman, D., Merril, C. R., and Bush, M. 1983. The cheetah is depauperate in genetic variation. Science 221:459–461.

Opdam, P. 1991. Metapopulation theory and habitat fragmentation: a review of holarctic breeding bird species. Landscape Ecology 5:93–106.

Paetkau, D. 2004. The optimal number of markers in genetic capture-mark-recapture studies. Journal of Wildlife Management 68:449–452.

Parker, W. T. 2006. Immobilization of small mammals and occupancy, seasonal food habits, and parasites of Allegheny woodrats in the Cumberland Mountains of Tennessee. M.S. Thesis, University of Tennessee, Knoxville, Tennessee.

Peakall, R., Ebert, D., Cunningham, R., and Lindenmayer, D. 2006. Mark-recapture by genetic tagging reveals restricted movements by bush rats (*Rattus fuscipes*) in a fragmented landscape. Journal of Zoology 268:207–216.

Poole, E. L. 1940. A life history sketch of the Allegheny woodrat. Journal of Mammalogy 21: 249–270.

Schlie, W. J. 1985. Reintroduction of the Allegheny woodrat (*Neotoma floridana magister*) to Neotoma Valley, Hocking County, Ohio. M.S. Thesis, Ohio State University, Columbus, Ohio.

Schlie, W. J., and Bookhout, T. A. 1985. Reintroduction of Allegheny woodrats to Hocking County, Ohio (abstract of meeting presentation). Ohio Journal of Science 85:92–93.

Scott, J. M., Davis, F., Csuti, B., Noss, R., Butterfield, B., Groves, C., Anderson, H., Caicco, S., D'Erchia, F., Edwards, T. C., Jr., Ulliman, J., and Wright, R. G. 1993. Gap analysis: a geographic approach to protection of biological diversity. Wildlife Monographs 123:1–41.

Serfass, T. L. 1998. River otter reintroduction: conflict in New York. IUCN Reintroduction Specialist Group Newsletter 16:11–12.

Serfass, T. L., Brooks, R. P., Rymon, L. M., and Rhodes, O. E., Jr. 2003. River otters in Pennsylvania, USA: lessons for predator reintroduction. *In* Proceedings of the European Otter Conference, eds. J. W. H. Conroy, A. C. Gutleb, J. Ruiz-Olmo and G. M. Yoxon. Broadford, Isle of Skye, Scotland.

Serfass, T. L., Novak, J. M., Johns, P. E., and Brooks, R. P. 1998. Genetic variation among river otter populations in North America: considerations for reintroduction projects. Journal of Mammalogy 79:736–746.

Serfass, T. L., Peper, R., Whary, M., and Brooks, R. P. 1993. River otter restoration in Pennsylvania: prerelease care and clinical evaluation. Journal of Zoo and Wildlife Medicine 24:28–40.

Serfass, T. L., Whary, M. T., Peper, R. L., Brooks, R. P., Swimley, T. J., Lawrence, W. R., and Rupprecht, C. E. 1995. Rabies in a river otter (*Lutra canadensis*) intended for reintroduction. Journal of Zoo and Wildlife Medicine 26:311–314.

Shafer, M. L. 1981. Minimum population sizes for species conservation. Bioscience 31:131–134.

Trefethen, J. B. 1975. An American crusade for wildlife. Winchester Press, New York.

Viggers, K. L., Lindenmayer, D. B., and Spratt, D. M. 1993. The importance of disease in reintroduction programmes. Wildlife Research 20:687–698.

Wells, J. V., and Richmond, M. E. 1995. Populations, metapopulations, and species populations: what are they and who should care? Wildlife Society Bulletin 23:458–462.

White, G. C. 2000. Population viability analysis: data requirements and essential analysis. *In* Research Techniques in Animal Ecology, eds. L. Boitani and T. K. Fuller, pp. 288–327. Columbia University Press, New York.

Wiens, J. A. 1976. Population responses to patchy environments. Annual Review of Ecology and Systematics 7:81–120.

Williams, C. L., Serfass, T. L., Cogan, R., and Rhodes, O. E., Jr. 2002. An extreme example of the genetic consequences of wildlife reintroductions: levels of microsatellite variation in the Pennsylvania elk herd. Molecular Ecology 11:1299–1310.

Wright, J. 1998. Radiotelemetry study of movements and resource use by Allegheny woodrats (*Neotoma magister*) in Pennsylvania. Project report to the Wild Resource Conservation Fund, Harrisburg, Pennsylvania.

Wright, S. 1969. Evolution and Genetics of Populations. Volume 2: The Theory of Gene Frequencies. University of Chicago Press, Chicago, Illinois.

Chapter 11
Status and Conservation of Other At-Risk Species of North American Woodrats

George A. Feldhamer and Aaron K. Poole

Introduction

Although the main focus of this volume centers on the Allegheny woodrat (*Neotoma magister*), there are 18 other species of woodrats in North America (Table 11.1) and several present serious conservation challenges. Some of these species, or subspecies, of woodrats have declined for many of the same reasons as *N. magister*, whereas the decline of others relates to unique factors associated with insular distribution. In this chapter, we summarize conservation efforts toward recovery of former population abundance and distribution of woodrat taxa that are listed at the federal or state level. We specifically focus on the conservation status of the state-endangered eastern woodrat (*N. floridana illinoensis*) in Illinois, and describe a translocation and reintroduction effort initiated there. Conservation and management of this species is discussed in the context of general principles that are applicable to the closely related *N. magister* and other declining species.

Current Woodrat Classification

Classification of species within the genus *Neotoma* has undergone substantial change during the past two decades. Wilson and Cole (2000) recognized 20 species of woodrats, 19 of which occur in North America. They considered the Turner Island woodrat a species (*N. varia*), but interestingly, did not recognize the Allegheny woodrat as a species distinct from the eastern woodrat. As part of their monumental re-evaluation of rodent classification based on morphological and molecular analyses, Musser and Carleton (2005) recognized 22 species within *Neotoma*. They considered the Turner Island woodrat a subspecies of white-throated woodrat (*N. albigula varia*) but accepted three species that were not recognized by Wilson

G.A. Feldhamer
Department of Zoology, Southern Illinois University, Carbondale, IL 62901

J.D. Peles, J. Wright (eds.), *The Allegheny Woodrat*,
DOI: 10.1007/978-0-387-36051-5_11,

Table 11.1 Species of *Neotoma* in North America[a], approximate number of recognized subspecies[b], geographic range, and current conservation status[c] (from Hall 1981, Monty and Emerson 2003, Musser and Carleton 2005)

Species	Number of Subspecies[d]	Geographic Range	Conservation Status
N. albigula; White-throated woodrat	11–15	SW United States and N Mexico	Least concern; Endangered as *N. varia*
N. angustapalata Tamaulipan woodrat	0	SW Tamaulipas, E Central Mexico	Least concern
N. anthonyi Anthony's woodrat	0	Todos Santos Islands, NW Baja, Mexico	Endangered; probably recently extinct
N. bryanti Bryant's woodrat	0	Cedros Island, W Central Baja, Mexico	Endangered, probably recently extinct
N. bunkeri Coronados Island woodrat	0	Coronados Island, S Baja, Mexico	Endangered, probably recently extinct
N. cinerea Bushy-tailed woodrat	13–16	NW Canada, W United States	Least concern
N. devia Arizona woodrat	0	W Arizona	Least concern
N. floridana Eastern woodrat	9–10	SE and Central United States	Least concern; Endangered as *N. f. smalli*
N. fuscipes Dusky-footed woodrat	8–11	Coastal W United States, N Baja, Mexico	Least concern; Critically endangered as *N. f. riparia*
N. goldmani Goldman's woodrat	0	SE Chihuahua and S San Luis Potosi, Mexico	Least concern
N. lepida Desert woodrat	28–31	W United States, Baja California	Least concern
N. leucodon White-toothed woodrat	8	W United States, N Mexico	Data deficient
N. macrotis Big-eared woodrat	7	Central California to N Baja, Mexico	Data deficient
N. magister Allegheny woodrat	0	East central United States	Near threatened, certainly declining throughout range
N. martinensis San Martin woodrat	0	San Martin Island, NW Baja, Mexico	Endangered, probably recently extinct
N. mexicana Mexican woodrat	24–26	W Central United States and Mexico	Least concern
N. micropus Southern plains woodrat	5	W Central United States, NE Mexico	Least concern
N. phenax Sonoran woodrat	0	NW Mexico	Least concern
N. stephensi Stephens woodrat	2	SW United States	Least concern

[a] North of the Tropic of Cancer

[b] Subspecies designations differ in Hall (1981) and Musser and Carleton (2005). Many species are in need of revision

[c] Classification from Baillie et al. (2004)

[d] For some species, such as *N. leucodon* and *N. macrotis*, "meaningful delimitation of subspecies is currently unwarranted" (Musser and Carleton 2005:1057).

and Cole (2000): (1) the white-toothed woodrat (*N. leucodon*), recently elevated to species status from *N. albigula* (Edwards et al. 2001); (2) the big-eared woodrat (*N. macrotis*; see Matocq 2003), formerly considered a subspecies of the dusky-footed woodrat (*N. fuscipes*); and (3) the Allegheny woodrat. There are two other species of woodrats in Mexico in the genus *Nelsonia*—called the diminutive woodrats. One of these species, the western diminutive woodrat (*N. neotomodon*), ranges into North America in southern Durango, Mexico.

Woodrats and Rarity

Most species worldwide, and most North American mammals, can be considered rare (Gaston 1994, 1997)—including woodrats. Likewise, most species have a relatively limited geographic range (Pagel et al. 1991). There are several ecological determinants that affect the rarity and potential for endangerment of woodrats and other species. In addition to geographic range, local densities and breadth of resources used are key factors in the relative abundance or rarity of a species (Rabinowitz 1981, Rabinowitz et al. 1986). Other key determinants include body size, interspecific competitors, reproductive potential, dispersal ability, and population genetics (see references in Feldhamer and Morzillo 2007). These factors, as well as extrinsic anthropogenic influences, act synergistically to impact several taxa of woodrats throughout North America. Several woodrat taxa are federally endangered as a result, and others are listed at the state level.

Federal and State-Listed Woodrat Species

Baja California Island Endemics

Four species of woodrats restricted to small islands off the coast of Baja California and derived from subspecies of the desert woodrat (*N. lepida*) are listed as endangered (Baillie et al. 2004). One presumptive subspecies of the white-throated woodrat (*N. albigula varia*), accorded species rank by some authorities as *N. varia*, also is endangered. In most cases, the islands on which the woodrats occur are very small. They were connected to the mainland by land bridges during the Pleistocene but have been separated for 7,500–10,000 years (Smith 1992, Hafner and Riddle 1997). Interestingly, these species of woodrats exhibit the "island rule"—a trend of insular small herbivores toward gigantism compared to the mainland taxa from which they diverged (Foster 1964). Conversely, large species trend toward dwarfism on islands.

All insular endemics have very limited geographic distributions and small population sizes, such that they are highly vulnerable to external impacts. For example, Anthony's woodrat (*N. anthonyi*) is restricted to Todos Santos Islands, approximately 14 km west of Ensenada in extreme northern Baja (Cortés-Calva et al. 2001).

Together, these two small islands comprise only about 130 ha in surface area, with vegetation characteristic of California coastal scrub (Wiggins 1980). The woodrat and deer mouse (*Peromyscus maniculatus*) represent the only two native terrestrial mammals on the islands. Anthony's woodrat apparently was abundant throughout the early 1900s (Nelson 1922). Sometime in the late 1980s or early 1990s, however, the species succumbed to predation by feral domestic cats (Mellink 1992) and is now believed to be extinct.

Bryant's woodrat (*N. bryanti*) is restricted to Cedros Island, which is 23 km northwest from the mainland. This island is about 360 km^2 in area with coastal, sand dune, and juniper (*Juniperus* spp.) scrub, chaparral, and pine (*Pinus* spp.) forest habitats. Woodrats were once widespread with nests found in a variety of plants and rock crevices (Alvarez-Castañeda and Yensen 1999). Unlike Todos Santos Islands, there are several native and introduced terrestrial mammals on Cedros Island, many that no doubt negatively impact Bryant's woodrat. Native mammals include endemic subspecies of the brush rabbit (*Sylvilagus bachmani*), San Diego pocket mouse (*Chaetodipus* [*Perognathus*] *fallax*), cactus mouse (*Peromyscus eremicus*), and mule deer (*Odocoileus hemionus*). Introduced species include feral cats and dogs, as well as domestic goats, sheep, cattle, and burros. Predation by cats and dogs, combined with loss of habitat from anthropogenic activities and grazing, has resulted in dramatic declines in the population density of Bryant's woodrat (Alvarez-Castañeda and Yensen 1999). The species, like *N. anthonyi*, probably is no longer extant.

The San Martin Island woodrat (*N. martinensis*) is another island endemic that is most likely extinct. It is restricted to the 318-ha island of San Martin in the Pacific Ocean approximately 5 km west of San Quintin, Mexico. Covered with lava beds that provide cracks and crevices for the woodrats, and coastal scrub vegetation, the island has no source of fresh water. The San Martin Island woodrat was considered numerous during the early 1900s. Following brush removal and introduction of feral cats by fishermen, however, the woodrat population declined significantly such that researchers in the 1960s no longer caught woodrats (Cortés-Calva et al. 2001).

Bunker's woodrat (*N. bunkeri*), restricted to Coronados Island, is known from only six specimens collected in 1931 (Smith 1992, Smith et al. 1993). At that time, however, it was considered to be "common" (Burt 1932). Coronados Island, about 8.5 km^2 in size, is in the Gulf of California about 2 km off southeastern Baja. This large woodrat is closely associated with the dark volcanic rock on the western part of the island. Essentially, the same factors associated with the decline of island endemics—feral cats introduced by fishermen and reduced habitat—have contributed to the probable extinction of this species (Smith et al. 1993, Alvarez-Castañeda and Ortega-Rubio 2003). The Turner Island woodrat (*N. albigula varia*) is considered endangered as well because of these same factors. This woodrat occurs only on Turner Island (also called Datil Island). Located in the Gulf of California, the island is only 2 km^2 in size. Efforts to document the continued occurrence of *N. a. varia* have been unsuccessful, and the taxon is most likely extinct.

Insular populations are especially vulnerable to extinction or decline. Therefore, Anthony's, Bryant's, San Martin Island, Bunker's, and Turner Island woodrats—all of which are likely extinct—may only foreshadow continued loss of woodrat taxa and other fauna on many other islands. There are more than 900 islands off Baja California with 28 endemic species and 220 subspecies of mammals (Alvarez-Castañeda and Ortega-Rubio 2003). As noted by Smith et al. (1993) "... increasing population pressure and greater accessibility have resulted in increased usage of the Gulf islands by both tourists and local fishermen." Although not currently listed, Alvarez-Castañeda and Ortega-Rubio (2003) considered *N. lepida latirostra* on Danzante Island (4.9 km^2) and *N. l. abbreviata* on San Francisco Island (2.6 km^2) among the most vulnerable taxa to predation by feral cats and resulting extinction. Management options suggested to address the threats to surviving island populations of woodrats include captive breeding programs, eradication of feral cats, reduced human access to islands with at-risk species, habitat restoration, and translocations of woodrats following cat removal and habitat restoration (Alvarez-Castañeda and Ortega-Rubio 2003).

The Riparian Woodrat

Another taxon adversely impacted by habitat loss and modification is the San Joaquin Valley or riparian woodrat, a subspecies of the dusky-footed woodrat (*N. fuscipes riparia*). Federally endangered since 2000 and considered a species of Special Concern in California, this large woodrat was first described by Hooper (1938). It has lighter pelage coloration, white rather than dusky feet, and more of a bicolored tail than other subspecies of *N. fuscipes*. Riparian woodrats are restricted to remnant areas of oak (*Quercus*) forested habitat with dense shrub cover about 100 ha in size along the Stanislaus River in Caswell Memorial State Park, Stanislaus County, California. This is believed to be a significant reduction in range from 70 years ago (Williams 1986). Population size in the park was estimated by Williams (1993) to be <450 individuals. The decline of this population is linked to several interacting factors, including the significant reduction and fragmentation of riparian habitat resulting from construction projects and water diversion, as well as trampling and grazing of these areas by cattle. Additionally, flooding by the Stanislaus River can destroy woodrat houses and indirectly impact the animals themselves, because the surrounding agricultural habitat provides no refuge. Populations are further impacted by rodenticide application, brush and tree removal, river channelization, and other anthropocentric activities. No conservation efforts are currently underway, however, specifically targeting the riparian woodrat (Hafner et al. 1998).

The Key Largo Woodrat

Recovery efforts for the Key Largo woodrat (*N. floridana smalli*) are probably the best documented of any woodrat taxon (Barbour and Humphrey 1982, McCleery et al. 2006). This is the southernmost of five subspecies of eastern woodrat that now occupies less than half its historic range. It is restricted to about 850 ha of tropical hardwood forest in the northern portion of Key Largo, Monroe County, Florida. Federally listed in 1984, the Key Largo woodrat population has continued to decline, despite the fact that most of the inhabited area is protected as part of Crocodile Lake National Refuge and Dagney Johnson Key Largo Hammock Botanical State Park. Habitat loss and fragmentation from development, feral cats, competition from invasive black rats (*Rattus rattus*) attracted by trash dumping, and fire ants (*Solenopsis* spp.) have all been proposed as factors effecting population decline. Quantification is lacking, however, as to the extent to which these factors have contributed to the decline of the Key Largo woodrat. A detailed recovery plan is now in place, but recent surveys suggest no more than 100 individuals remain. Although captive breeding and subsequent release have been proposed as part of the recovery process, projections from modeling suggest that "... captive breeding and supplementation of the wild population ... will have minimal long term impacts on the recovery" of the population (McCleery et al. 2005). Unfortunately, there is uncertainty over the best management directions, and as noted by McCleery et al. (2006), "To date, descriptions of [Key Largo woodrat] habitat use and nesting ecology have been contrary and confusing ..."

Woodrats in Southern Illinois

Background

The eastern woodrat is listed as state-endangered in Illinois (Herkert 1992). Historically, the species was found throughout southern Illinois, primarily within Shawnee National Forest, but has been extirpated throughout most of this area in the 1900s (Nawrot and Klimstra 1976). The reasons for this decline are unclear. Nawrot and Klimstra (1976) cited two harsh winters during the early 1900s as a possible reason. It seems unlikely, however, that two bad winters nearly 100 years ago would have such a long-term impact on the current population. Periodic harsh winters could negatively affect populations already declining from other factors, however. In addition to climate, factors that could negatively impact the woodrat population in Illinois are the same as those that can affect the Allegheny woodrat. These include decreased habitat quality (for example, reduced mast production), habitat fragmentation, predation, or competition. Parasitism by the nematode *Baylisascaris procyonis* carried by raccoons (*Procyon lotor*), may also be a factor. Birch et al. (1994) found only 3 of 60 raccoons (5.0%) from southern Illinois carried *B. procyonis*, and suggested that it was a negligible factor in woodrat declines. Nonetheless, *B. procyonis* could be more prevalent than previously suspected. Habitat fragmentation is apparent throughout Shawnee National Forest, but the impacts of other factors on woodrats remain unknown.

Monty et al. (1995) discovered previously unknown woodrat populations in southwestern Illinois by investigating areas where they occurred historically but were believed to be extirpated. Prior to these discoveries, the distribution of woodrats

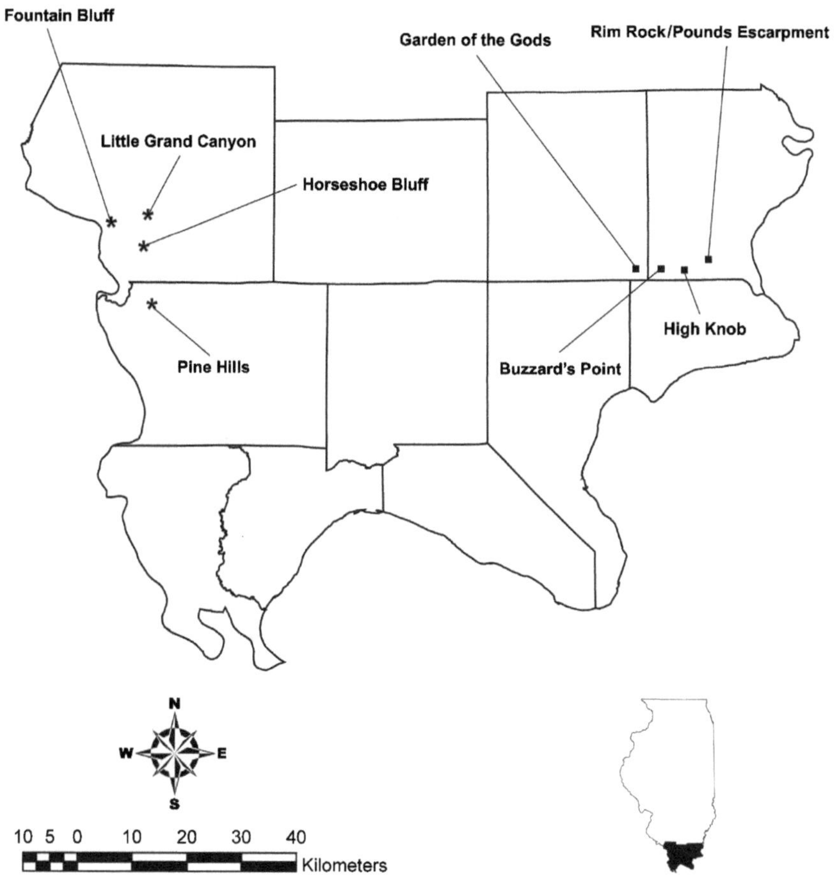

Fig. 11.1 Location in southern Illinois of naturally occurring populations (*) of eastern woodrats (*N. f. illinoensis*) and translocated populations (■)

in Illinois was thought to consist of only one population located at La Rue-Pine Hills, Jackson County. Using microsatellite DNA at six loci, evaluation of the genetic diversity of these recently discovered populations at Fountain Bluff, Little Grand Canyon, and Horseshoe Bluff (Fig. 11.1), as well as the known population at Pine Hills, showed very low heterozygosity and low proportions of polymorphic loci (Monty et al. 2003). This indicated high levels of inbreeding with little exchange of genetic material between populations despite the fact that Little Grand Canyon, Horseshoe Bluff, and La Rue-Pine Hills form a north–south continuum separated by only about 2 km with limited areas of unsuitable habitat between them.

Reintroduction Efforts

A collaborative effort by biologists from Illinois Department of Natural Resources, United States Forest Service, and Southern Illinois University Carbondale was initiated to re-establish and recover woodrats in southern Illinois. It was determined that this could best be accomplished by translocating animals to formerly occupied areas in the eastern portion of the Shawnee National Forest. Prior to reintroductions, Novosak (2004) identified suitable release sites from among formerly occupied locations based on a habitat suitability index (HSI) derived for the area. Habitat variables used were diversity and density of mast producing trees, density of understory vegetation, and extent and quality of rocky outcrops in terms of ledges, cracks, depressions, and associated features. Two locations (Garden of the Gods, Saline County, and Rim Rock/Pounds Escarpment, Gallatin County) approximately 9.5 km apart were chosen based on HSI values. Translocations were initiated in April 2003 with donor *N. f. illinoensis* from sites in Arkansas and Missouri. Translocations began at High Knob, which is approximately equidistant between the initial two release sites (Fig. 11.1), in October 2005. An additional reintroduction of woodrats was made to Buzzard's Point—equidistant between Garden of the Gods and High Knob—in February and March 2007.

From April 2003 through March 2007, 341 woodrats were translocated from various locations in Missouri, Arkansas, and the LaRue-Pine Hills population to the four locations in southeastern Illinois. A "soft" release technique was used, with artificial stick nests containing food and bedding material constructed for each woodrat prior to its release. Survival and extent of reproduction were estimated through periodic mark and recapture monitoring at points of release and the surrounding areas.

Radiotelemetry was used to evaluate movement patterns at the Garden of the Gods and Rim Rock/Pounds Escarpment sites. Radio transmitters were placed on 15 individuals at Garden of the Gods. Only six individuals could be located one week after release. Of those, only one animal was recaptured. Two transmitters were found with the remains of the woodrats. Radio transmitters were placed on six individuals released at Rim Rock/Pounds Escarpment. All animals were living two days after release, but after four days, only four individuals could be located. Two transmitters probably slipped off the animals. Signals were detected throughout the life of the remaining transmitters, and nightly movements between the rocky outcrops and the surrounding forest were determined, but no animals were recaptured.

Preliminary population estimates were very difficult to establish because of low numbers of recaptured animals. Many released animals may have successfully dispersed from the areas. Radiotelemetry results were inconclusive because inability to relocate animals may have resulted from faulty transmitters, insufficient signal strength of transmitters from animals that immediately moved out the release area, predation that destroyed the transmitter, or natural mortality of woodrats in rocky habitat that blocked transmitter signals. Mark and recapture seems to be the most effective way of determining the population status of translocated woodrats in southern Illinois. Of the 341 individuals translocated to previously occupied areas, at least 30 (8.8%) were known to be alive $\geq$1 month after release. More encouraging is the fact that reproduction was documented, based on live-trapping of 58 unmarked individuals among the four release sites. Whether these 58 woodrats were offspring of pregnant females released or the result of reproduction occurring on site will be determined from microsatellite DNA obtained from pinnae tissue of each woodrat captured.

Novosak (2004) translocated 12 woodrats from Missouri to Pine Hills in southwestern Illinois in an attempt to increase genetic diversity of that population. In February 2005, 29 additional woodrats from Arkansas were translocated to Pine Hills. In a similar manner, six woodrats from Missouri were translocated to Fountain Bluff during April 2005 in an initial attempt to increase the genetic diversity of that population.

Conservation and Management of Declining Woodrat Species

Reasons for Decline

Whether we consider insular species of woodrats, or mainland populations such as the riparian woodrat or the eastern woodrat in southern Illinois, Diamond (1989) and Caughley and Gunn (1996) noted four general factors that can lead to extirpation or extinction of species: (1) overharvest; (2) habitat modification and fragmentation; (3) introduced species; and (4) secondary extinctions (where extinction of one taxon predisposes other taxa to extinction). Insular species are particularly susceptible to extinction (Primack 2000, Simberloff 2000). Occupied areas often are limited, and small populations can quickly succumb to various stochastic or anthropogenic factors, as has occurred with woodrats on islands of Baja California and Key Largo.

For mainland woodrat populations, habitat modification and fragmentation, and possibly introductions, appear to be the primary factors in population declines, although quantitative data are often lacking. Several other site-specific factors (predation, weather, unpredictable food resources, parasites, and disease) most likely act synergistically to impact mainland woodrat populations spatially and temporally as well (LoGiudice 2006). Small population sizes and resultant declines may be exacerbated by the basic biology of woodrats. Many species—including the eastern woodrat—are larder hoarders. They cache mast or other food in one or few locations. As a result, individuals are generally intolerant of conspecifics and are highly territorial of nest sites and caches (Fitch and Rainey 1956, Rainey 1956, Kinsey 1977). Juveniles, especially males, often are driven out of the area. Population sex ratios may be highly skewed toward females (Monty 1997) and the effective population size reduced. This leads to the potential for genetic drift and inbreeding depression—again, exemplified by the eastern woodrat in southern Illinois.

Additionally, the "packrat" behavior of woodrats may be a causative factor in population decline where population densities of raccoons are high. As noted, raccoons can carry the nematode parasite *B. procyonis*, with eggs shed in the feces. Like *N. magister*, eastern woodrats probably carry raccoon feces to their food caches. They then remove seeds from the feces, but only several weeks later when the parasite eggs have become infective (LoGiudice 2001) and highly pathogenic to woodrats.

Recovery and Management

Just as population declines of Allegheny and eastern woodrats share many causal factors, there are commonalities in conservation strategies for population recovery and management, habitat management, as well as human dimensions.

Population Management

Live-trapping should be conducted at historic sites where woodrats were known to occur—based on previous literature or evidence of abandoned houses—and at potentially active sites where adequate habitat currently exists. If no baseline data exist for the study site, it is necessary to determine the current distribution and population abundance of extant woodrats. At sites where woodrats occur, a long-term monitoring protocol should be established to determine population demographics. Protection as a state-listed species should be considered in coordination with state and federal regulatory agencies, if justified based on reduced site occupancy and declining population trends. Protection may involve reducing human disturbance and potential mortality from predators. Reduction of *B. procyonis* can be attempted through controlling raccoon populations (LoGiudice 2003) or use of antihelmintic baits.

In addition to protecting existing populations, it may be desirable to augment them with individuals translocated from other areas, or to reintroduce woodrats to previously occupied sites where suitable habitat is available. Success of reintroduction is usually defined as establishing a viable, self-sustaining population (see Beck et al. 1994). There has been an exponential increase in the number of reintroduction efforts worldwide since the 1980s (Seddon et al. 2007). Formation of the Reintroduction Specialist Group of the Species Survival Commission (IUCN) as well as the Reintroduction Advisory Group of the American Association of Zoological Parks and Aquariums reflects the increased interest in reintroductions. Based on our efforts with eastern woodrats, we suggest the recommendations discussed by Serfass (Chapter 10) with additional consideration of the following:

(1) Obtain animals from as many donor populations as possible to increase the number of individuals available and enhance the genetic diversity of the reintroduced population(s);
(2) Release in accessible sites to facilitate efficient monitoring by management personnel involved in time and labor-intensive live-trapping;
(3) Hold donor animals at the trap site for the minimum time possible;
(4) Provide a "soft" release (Caughley and Gunn 1996) with a ready-made cache of acorns or other suitable food and a box with nesting materials, covered by appropriate sized sticks to mimic a natural nest. Construct at intervals throughout the release site prior to bringing in animals to minimize the time woodrats are held.

Habitat Management

GIS databases and GAP analysis for the study area can provide data on habitat for existing or reintroduced populations. Habitat restoration of woodlands and riparian areas on public lands can be coordinated with state agencies, especially if the woodrat is a state-listed species. Reduce exotic vegetation if it negatively impacts native food resources or cover for woodrats. Additional habitat—especially potential corridors and contiguous areas that will reduce fragmentation—may be integrated with large-scale multiple use initiatives of management agencies to benefit woodrats. Habitat acquisition of private lands is often problematic and financially prohibitive. Although conceptually straightforward, these types of efforts require establishing partnerships between resource agencies, the general public, and private landowners—often difficult to do in practice—but essential if woodrats are to be managed as a metapopulation.

Human Dimensions

In addition to the organizational aspects of species recovery (Clark and Cragun 1994), successful population and habitat management necessitates increased public awareness, empathy, and stewardship toward woodrats. We have found that the common name "woodrat" is often perceived as synonymous with "disease-ridden sewer rats" among the general public. The usual question is "why do we want to save rats?" Biologists must help dispel this unfortunate public image whenever possible through discussions with landowners, presentations to local conservation organizations and school groups of all age levels, and in conjunction with state resource agencies developing educational materials for schools, state parks, and information centers.

Our woodrat reintroduction program remains ongoing with the ultimate outcome to be determined. Habitat fragmentation remains a factor. Other factors that may have contributed to the past extirpation of the eastern woodrat throughout southern Illinois may still be operative as well, including unpredictable food supply, predation, extreme weather, or mortality from *B. procyonis*. Nonetheless, our reintroduction effort appears to be successful in terms of existing population sizes, natural reproduction, and dispersal into areas ≥ 1 km from release sites. Whether this initial success continues longer term will be determined by continued population monitoring.

References

Alvarez-Castañeda, S. T., and Ortega-Rubio, A. 2003. Current status of rodents on islands in the Gulf of California. Biological Conservation 109:157–163.

Alvarez-Castañeda, S. T., and Yensen, E. 1999. *Neotoma bryanti*. Mammalian Species 619:1–3.

Baillie, J. E. M., Hilton-Taylor, C., and Stuart, S. N. (eds). 2004. 2004 IUCN Red List of Threatened Species. A Global Species Assessment. IUCN, Gland, Switzerland and Cambridge, United Kingdom.

Barbour, D. B., and Humphrey, S. R. 1982. Status and habitat of Key Largo woodrat and cotton mouse (*Neotoma floridana smalli* and *Peromyscus gossypinus allapaticola*). Journal of Mammalogy 63:144–148.

Beck, B. B., Rapaport, L. G., Stanley Price, M. R., and Wilson, A. C. 1994. Reintroduction of captive-born animals. *In* Creative Conservation: Interactive Management of Wild and Captive Animals, eds. P. J. S. Olney, G. M. Mace and A. T. C. Feistner, pp. 265–286. Chapman and Hall, New York.

Birch, G. L., Feldhamer, G. A., and Dyer, W. G. 1994. Helminths of the gastrointestinal tract of raccoons in southern Illinois with management implications of *Baylisascaris procyonis* occurrence. Transactions of the Illinois State Academy of Science 87:165–170.

Burt, W. H. 1932. Description of heretofore unknown mammals from islands in the Gulf of California, Mexico. Transactions of the San Diego Society of Natural History 7:161–182.

Caughley, G., and Gunn, A. 1996. Conservation Biology in Theory and Practice. Blackwell Science, Cambridge, Massachusetts.

Clark, T. W., and Cragun, J. R. 1994. Organizational and managerial guidelines for endangered species restoration programs and recovery teams. *In* Restoration of Endangered Species, eds. M. L. Bowles and C. J. Whelan, pp. 9–33. Cambridge University Press, New York.

Cortés-Calva, P., Alvarez-Castañeda, S. T., and Yensen, E. 2001. *Neotoma anthonyi*. Mammalian Species 663:1–3.

Cortés-Calva, P., Yensen, E., and Alvarez-Castañeda, S. T. 2001. *Neotoma martinensis*. Mammalian Species 657:1–3.

Diamond, J. 1989. Overview of recent extinctions. *In* Conservation for the Twenty-first Century, eds. D. Western and M. C. Pearl, pp. 37–41. Oxford University Press, New York.

Edwards, C. W., Fulhorst, C. F., and Bradley, R. D. 2001. Molecular phylogenetics of the *Neotoma albigula* species group: Further evidence of a paraphyletic assemblage. Journal of Mammalogy 82:267–279.

Feldhamer, G. A., and Morzillo, A. T. 2007. Relative abundance and conservation: is the golden mouse a rare species? *In* The Golden Mouse: Ecology, Behavior, and Conservation, eds. G. W. Barrett and G. A. Feldhamer, pp. 113–129. Springer-Verlag, New York.

Fitch, H. S., and Rainey, D. G. 1956. Ecological observations on the woodrat, *Neotoma floridana*. University of Kansas Museum of Natural History Publications 8:499–533.

Foster, J. B. 1964. Evolution of mammals on islands. Nature 202:234–235.

Gaston, K. J. 1994. Rarity. Chapman and Hall, New York.

Gaston, K. J. 1997. What is rarity? *In* The Biology of Rarity, eds. W. E. Kunin and K. J. Gaston, pp. 30–47. Chapman and Hall, London, United Kingdom.

Hafner, D. J., and Riddle, B. R. 1997. Biogeography of Baja California peninsular desert mammals. *In* Life Among the Muses: Papers in Honor of James S. Findley, eds. T. L. Yates, W. L. Gannon and D. E. Wilson, pp. 39–68. Museum of Southwestern Biology, University of New Mexico, Albuquerque, New Mexico.

Hafner, D.J., Yensen, E., and Kirkland, Jr., G.L. (eds.). 1998. North American Rodents: Status Survey and Conservation Action Plan. IUCN/SSC Rodent Specialist Group, Gland, Switzerland.

Hall, E. R. 1981. The mammals of North America. Second edition. John Wiley and Sons, New York.

Herkert, J. R. (ed.). 1992. Endangered and Threatened Species of Illinois: Status and Distribution. Volume 2: Animals. Illinois Endangered Species Protection Board, Springfield, Illinois.

Hooper, E. T. 1938. Geographical variation in wood rats of the species *Neotoma fuscipes*. University of California Publications in Zoology 42:213–246.

Kleiman, D. G., Stanley Price, M. R., and Beck, B. B. 1994. Criteria for reintroductions. *In* Creative Conservation: Interactive Management of Wild and Captive Animals, eds. P. J. S. Olney, G. M. Mace, and A. T. C. Feistner, pp. 287–303. Chapman and Hall, New York.

Kinsey, K. P. 1977. Agonistic behavior and social organization in a reproductive population of Allegheny woodrats, *Neotoma floridana magister*. Journal of Mammalogy 58:417–419.

LoGiudice, K. 2001. Latrine foraging strategies of two small mammals: implications for the transmission of *Baylisascaris procyonis*. American Midland Naturalist 146:369–378.
LoGiudice, K. 2003. Trophically transmitted parasites and the conservation of small populations: Raccoon roundworm and the imperiled Allegheny woodrat. Conservation Biology 17: 258–266.
LoGiudice, K. 2006. Toward a synthetic view of extinction: a history lesson from a North American rodent. BioScience 56:687–693.
Matocq, M. D. 2003. Morphological and molecular analysis of a contact zone in the *Neotoma fuscipes* species complex. Journal of Mammalogy 83:866–883.
McCleery, R. A., Lopez, R. R., Silvy, N. J., and Grant, W. E. 2005. Effectiveness of supplemental stockings for the endangered Key Largo woodrat. Biological Conservation 124:27–33.
McCleery, R. A., Lopez, R. R., Silvy, N. J., Frank, P. A., and Klett, S. B. 2006. Population status and habitat selection of the endangered Key Largo woodrat. American Midland Naturalist 155:197–209.
Mellink, E. 1992. The status of *Neotoma anthonyi* (Rodentia, Muridae, Cricetinae) of Todos Santos Islands, Baja California, Mexico. Bulletin of the Southern California Academy of Science 91:137–140.
Monty, A-M. 1997. The eastern woodrat (*Neotoma floridana*) in southern Illinois: population assessment and genetic variation. Ph.D. Dissertation, Southern Illinois University, Carbondale.
Monty, A-M., and Emerson, R. E. 2003. Eastern woodrat, *Neotoma floridana*, and allies. *In* Wild Mammals of North America: Biology, Management, and Conservation, eds. G. A. Feldhamer, B. C. Thompson, and J. A. Chapman, pp. 381–393. Johns Hopkins University Press, Baltimore, Maryland.
Monty, A-M., Wagle, E. R., Emerson, R. E., and Feldhamer, G. A. 1995. Recently discovered populations of eastern woodrats (*Neotoma floridana*) in southern Illinois. Transactions of the Illinois State Academy of Science 88:43–47.
Monty, A-M., Heist, E. J, Wagle, E. R., Emerson, R. E., Nicholson, E. H., and Feldhamer, G. A. 2003. Genetic variation and population assessment of eastern woodrats in southern Illinois. Southeastern Naturalist 2:243–260.
Musser, G. G., and Carleton, M. D. 2005. Superfamily Muroidea. *In* Mammal Species of the World: A Taxonomic and Geographic Reference. Second edition. D. E. Wilson and D. M. Reeder, pp. 894–1531. Johns Hopkins University Press, Baltimore, Maryland.
Nawrot, J. R., and Klimstra, W. D. 1976. Present and past distribution of the endangered southern Illinois woodrat (*Neotoma floridana illinoensis*). Chicago Academy of Science Natural History Miscellanea No. 196.
Nelson, E. W. 1922. Lower California and its natural resources. Memoirs of the National Academy of Science 16:1–194.
Novosak, B. A. 2004. Reintroduction of the eastern woodrat (*Neotoma floridana*) in southern Illinois: habitat assessment and release. M.S. Thesis, Southern Illinois University, Carbondale.
Pagel, M. D., May, R. M., and Collie, A. R. 1991. Ecological aspects of the geographic distribution and diversity of mammalian species. American Naturalist 137:791–815.
Primack, R. B. 2000. A primer of conservation biology. Second edition. Sinauer Associates, Sunderland, Massachusetts.
Rabinowitz, D. 1981. Seven forms of rarity. *In* The Biological Aspects of Rare Plant Conservation, ed. H. Synge, pp. 205–217. John Wiley and Sons, New York.
Rabinowitz, D., Cairns, S., and Dillon, T. 1986. Seven forms of rarity and their frequency in the flora of the British Isles. *In* Conservation Biology, the Science of Scarcity and Diversity, ed. M. E. Soulé, pp. 182–204. Sinauer, Sunderland, Massachusetts.
Rainey, D. G. 1956. Eastern woodrat, *Neotoma floridana*: Life history and ecology. University of Kansas Publications of the Museum of Natural History 8:535–646.
Seddon, P. J., Armstrong, D. P., and Maloney, R. F. 2007. Developing the science of reintroduction biology. Conservation Biology 21:303–312.

Simberloff, D. 2000. Extinction-proneness of island species – causes and management implications. The Raffles Bulletin of Zoology 48:1–9.
Smith, F. A. 1992. Evolution of body size among woodrats from Baja California, Mexico. Functional Ecology 6:265–273.
Smith, F. A., Bestelmeyer, B. T., Biardi, J., and Strong, M. 1993. Anthropogenic extinction of the endemic woodrat, *Neotoma bunkeri* Burt. Biodiversity Letters 1:149–155.
Wiggins, I. L. 1980. Flora of Baja California. Stanford University Press, Stanford, California.
Williams, D. F. 1986. Mammalian species of special concern in California. California Department of Fish and Game Report 86-1:1–112.
Williams, D. F. 1993. Population censuses of riparian brush rabbits and riparian woodrats at Caswell Memorial State Park during January 1993. Final report to the California Department of Parks and Recreation, Sacramento, California.
Wilson, D. E., and Cole, F. R. 2000. Common names of mammals of the world. Smithsonian Institution Press, Washington, D.C.

Section IV
Synthesis

Our initial impetus for compiling this book was to gather and organize the heretofore scattered results of more than two decades of research on the Allegheny woodrat (*Neotoma magister*). Like the many researchers whose work is cited in this book, we have found the intriguing habits and mysterious decline of this cryptic animal a fascinating and frustrating puzzle. Realizing that this animal is listed as threatened, endangered, extirpated, or a species of concern in more states than any other rodent, we thought it deserved special attention. We quickly realized, however, that the lessons learned and recommendations developed for the Allegheny woodrat can be a valuable template for many other species, particularly of small mammals, which have now got the attention of biologists and conservationists. In this section, as we consider what we have learned about investigating decline, understanding ecology and genetics, and planning for conservation, we recap those lessons and recommendations as guidelines for future research and action.

John D. Peles
Janet Wright

Chapter 12
Synthesis: Lessons Learned and Recommendations for the Future

John D. Peles and Janet Wright

Introduction

What began nearly 30 years ago with anecdotal observations concerning the absence of the Allegheny woodrat (*Neotoma magister*) from some historically occupied areas has resulted in the emergence of this species as a conservation concern because of declining numbers throughout a large portion of its range (Hicks 1989, Beans 1992). The chapters in this book represent the efforts of researchers and managers who have attempted to understand the reasons for the decline and, at the same time, develop a strategy for conservation and management of the species. Although much has been gained from these efforts, this body of work also illustrates the difficulties faced in the study of a small-sized, inconspicuous, patchily distributed species that is in decline.

Despite historical obstacles, we argue that much has been gained from the study of the Allegheny woodrat. In fact, this book is based on the premise that *N. magister* represents a model species for studying the general phenomenon of decline. The chapters in this book represent our current knowledge and provide insights into approaches for conservation and management that are applicable to the Allegheny woodrat as well as other declining species. This is especially true with the recognition that declines are occurring in other species or subspecies of *Neotoma* (Feldhamer and Poole Chapter 11).

In this chapter, our goal is to highlight important findings, in the form of lessons learned, from nearly 30 years of research on the Allegheny woodrat, without becoming lost in the details. These will be summarized in the areas of (1) the study of decline; (2) ecology; (3) genetics; and (4) tools for conservation and management. We synthesize findings in the context of conserving and managing declining populations of this species, but many will apply to other species as well. In addition, we make recommendations for future research, conservation, and management efforts focusing on the Allegheny woodrat.

J.D. Peles
Pennsylvania State University, Greater Allegheny, McKeesport, PA 15132

J.D. Peles, J. Wright (eds.), *The Allegheny Woodrat*,
DOI: 10.1007/978-0-387-36051-5_12,

Decline of the Allegheny Woodrat

Lessons Learned

The decline of the Allegheny woodrat went largely unrecognized for an extended period of time—Undoubtedly, one reason the decline was not recognized until the late 1970s was the classification of the Allegheny woodrat as a subspecies of *N. floridana* (Wright Chapter 1). Little attention was given to the biology of a subspecies, and any observations of locally declining numbers were likely viewed as insignificant in the context of a wide-ranging species such as *N. floridana.* Taxonomic revisions reflecting the recognition of subspecies as distinct species are becoming more common (Isaac et al. 2004). Thus, it is likely that other distinct, but not yet recognized, species are undergoing undetected declines.

Another factor that likely delayed recognition of the need for research and monitoring efforts for the Allegheny woodrat is that the general public has little knowledge of, or interest in, woodrats. In fact, some have suggested that the name "woodrat" may even inspire a negative public reaction (Hassinger et al. Chapter 8, Feldhamer and Poole Chapter 11). Rodents, in general, lack the charismatic appeal that attracts attention from conservationists. Ceballos and Brown (1995) noted that the majority of mammals appearing on the endangered species list in 1994 were highly visible species such as primates, ungulates, and carnivores. In contrast, rodents and bats (i.e., inconspicuous, noncharismatic species) represented most of the mammalian species known to have gone extinct in recent years (Ceballos and Brown 1995).

Lack of baseline data on distribution and abundance of the Allegheny woodrat made detection of the decline and determination of its extent difficult—The Allegheny woodrat is an inconspicuous species with unique habitat requirements that make it difficult to study. This, in addition to its taxonomic status, represents another likely reason that *N. magister* historically was given little attention by biologists. This lack of attention not only allowed the decline of *N. magister* to remain undetected until recently, but also has made it difficult to assess the current status of the species. Distributions within each state were poorly documented and there were no baseline population data available. After the initiation of statewide monitoring programs in the mid-1980s (Mengak et al. Chapter 7), declines have been documented in many states (Wright Chapter 1). However, the timing of the decline, as well as its extent compared to historic numbers and distribution, will almost certainly remain unknown. Many species share, to some degree, the same impediments to detection of status. For example, Martin et al. (2003) concluded that, although Franklin's ground squirrel (*Spermophilus franklinii*) is declining in Illinois, its status is difficult to determine because of its secretive nature and lack of knowledge concerning its biology.

A single-factor approach hindered our understanding of the reasons for the decline of the Allegheny woodrat—In a review of the literature, LoGiudice (2006) concluded that most investigations of species declines between 1996 and 2006 considered only a single factor. For nearly 20 years, this single-factor approach was used by biologists in studying the decline of the Allegheny woodrat. Although we will never know for certain the exact reasons for the demise of the woodrat, the decline of this species was clearly a complex process involving multiple factors (LoGiudice 2006, LoGiudice Chapter 2). Although others have recognized the importance of considering the synergistic interactions of multiple causes for species extinctions (Pimm 1996), the story of the Allegheny woodrat demonstrates another level of complexity in this process. It appears that different factors acted at different times, and in different ways (e.g., synergistic, additive), to cause the decline. Furthermore, certain factors may only have been important in certain areas (LoGiudice Chapter 2). Thus, future attempts to understand the decline of a species will do well to consider multiple causes within a temporal and spatial framework.

Factors currently influencing the Allegheny woodrat likely differ from those that initially caused its decline—In the context of conservation and recovery, it is evident that addressing factors currently affecting woodrat populations must receive a higher priority than understanding those that initially caused its decline. For example, although the raccoon roundworm (*Baylisascaris procyonis*) likely did not contribute to the initial decline of woodrats, this parasite is now prevalent in some areas (LoGiudice Chapter 2). Belovsky et al. (1994) have made the case that management efforts should focus not only on addressing the factor(s) believed to have caused the decline of populations, but also should consider those stressors presently acting on the population that might prevent its recovery. In the case of the Allegheny woodrat, these include three categories of factors. Some factors, such as reduced availability of mast (Castleberry and Castleberry Chapter 6) or timber harvesting (Castleberry Chapter 4), result directly or indirectly from human activity. Others, including increased demographic stochasticity and reduced genetic variation (Smyser and Rhodes Chapter 9) result from reduced population size. Lastly, population recovery might be hindered by factors, such as agonistic behavior and den use (Peles and Wright Chapter 5), that are inherent to the biology of the species.

Recommendations

Focus research efforts on understanding the factors currently influencing survivorship and reproductive success in the Allegheny woodrat—The decline of the Allegheny woodrat was initiated by the confluence of several historical factors and has been in progress for a relatively long time (i.e., at least 30 years). Because it is likely not possible to reverse this process, efforts should be focused on understanding those factors currently influencing the species to achieve the optimal current conditions for fostering recovery. Specifically, we suggest that future research should examine: (1) the influence of food quality and availability by providing mast supplementation; (2) the effects of antihelmintic drugs in areas where woodrats are at high risk to *B. procyonis* infection (Hassinger et al. Chapter 8); and (3) the effects of increased home range and foraging movements in response to timber harvesting or other disturbances. Integration of results from studies designed to answer these questions—along with consideration of factors such as genetic variation and behavioral characteristics—will be required to achieve the goals of conserving remaining woodrat populations and maximizing the probability of success for reintroduction efforts.

Use the current status of the Allegheny woodrat in each state as a basis for determining the role of that state in a comprehensive management strategy—Because *N. magister* is not federally listed, conservation and management decisions have historically been left to each state. Hassinger et al. (Chapter 8) have emphasized the need for coordination of management efforts among states in the form of a regional approach. However, administration of these efforts will ultimately be the responsibility of management agencies within individual states. Therefore, it is important for these agencies to recognize their role in a large-scale management strategy by directing conservation efforts appropriately based on the status of the species in that state. For those states that commit to action, these roles could include (1) reintroducing woodrats to those states where they have been extirpated; (2) implementing conservation strategies in states where they are declining; and (3) monitoring to detect possible declines in states where the species is stable.

Ecology of the Allegheny Woodrat

Lessons Learned

Allegheny woodrats exhibit a natural metapopulation structure—Nearly all of our chapter authors, whatever their focus, have referred to *N. magister* as an apparent metapopulation species. Classically, a metapopulation consists of subpopulations, some of which experience occasional failure; yet, the overall metapopulation persists through dispersal and recolonization among its subunits (Hanski 1996). The evidence of this condition for the Allegheny woodrat is a combination of data and logic. The existence of *N. magister's* preferred habitat in small, discrete patches over the landscape predisposes it to such a structure (Castleberry Chapter 4, Hassinger et al. Chapter 8). Genetic data showing a scale of differentiation among individual rock outcrops that is consistent with isolation by distance and small-scale gene flow (Wood Chapter 3, Smyser and Rhodes Chapter 9) bolster the metapopulation model, as does the repeated observation that apparently vacated habitat patches become recolonized (Mengak et al. Chapter 7). Although direct observation of dispersal events to connect patches is largely lacking, the circumstantial evidence is compelling (Wood Chapter 3).

Many mammalian species have been characterized as functioning at a metapopulation level since the concept was introduced, including large carnivores such as cougars (*Puma concolor*; Sweanor et al. 2000), social primates (Marsh and Wagner 2003) and especially small mammals (Krohne 1997, Telfer et al. 2001, Ozgul et al. 2006). A number of species—such as forest mammals in South Africa (Lawes et al. 2000) or prairie dogs (*Cynomys ludovicianus*) in the American west (Roach et al. 2001)—have recently been forced into a metapopulation mode as their natural habitats become fragmented by human activity. The Allegheny woodrat is unusual because it appears to be one of few mammalian species that exist in a natural metapopulation organization. Talus-dwelling pikas (*Ochotona princeps* and *Ochotona collaris*) in western North America (Moilanen et al. 1998, Franken and Hik 2004) are probably the best-demonstrated example. Other habitat specialists such as rock hyraxes (Hyracoidea; Gerlach and Hoeck 2001) and the swamp-dwelling quokka (*Setonix brachyurus*; Hayward et al. 2006) also represent species that likely exhibit a natural metapopulation structure. An implication of such structure, and one that eluded us for years, is that the sizes we typically record for Allegheny woodrat populations at an individual rock outcrop are, according to conventional wisdom, "impossibly" small. Often consisting of fewer than 20 individuals even in areas where no decline is evident (Wood Chapter 3, Mengak et al. Chapter 7), these population units exist despite stochastic fluctuations, high mortality rates and, for a rodent, low reproductive rates (Wood Chapter 3). The logical deduction is that since its emergence as a distinct species, roughly at the Pleistocene–Holocene transition some 15,000 years ago (Hayes and Harrison 1992), the Allegheny woodrat has defied the minimum population odds with behavioral adaptations that allowed frequent and successful colonization forays between rock outcrops within a metapopulation. A second implication, and the primary one for conservation, is that the same population organization that has succeeded for so long is now in collapse as a result of human activities. Understanding the population structure of the Allegheny woodrat turns out to be central to every consideration in interpreting its ecology and guiding its conservation.

Behaviors of the Allegheny woodrat, especially differential behaviors of males and females, determine how space is occupied—In the early years, Allegheny woodrat researchers informally referred to inhabitants of a single-rock patch as a "colony," but we have dropped this term as it has become apparent that social organization of the type evident in marmots (e.g., Frase and Hoffman 1980) or prairie dogs (e.g., Hoogland 1996) does not apply to this species. Observations in captivity indicate that *N. magister* is intolerant of neighbors (Kinsey 1976, 1977), and radiotelemetry studies suggest that adult females pre-empt favored den sites in the central rock patch, with males occupying more peripheral den sites (Peles and Wright Chapter 5). This may place a premium on the availability of high-quality dens at a rock outcrop, even when not all possible dens are occupied, so that behavior places further limitations on an already limiting resource. Males occupy larger home ranges than females (Castleberry Chapter 4), and are more likely to make long-range movements, whether to locate mates or to disperse to new habitat (Wood Chapter 3, Castleberry Chapter 4). By contrast, females are relatively philopatric, and juvenile females are more likely than juvenile males to remain near the natal site (Wood Chapter 3). If, as the limited evidence suggests, females are less likely than males to disperse to a new rock patch, this would increase the time needed for a local population to be re-established once it has disappeared. In the past, a "rescue effect" (Gonzalez et al. 1998)—wherein an occasional migrant male rejuvenated a critically low population—may have been more common than founders repopulating totally vacant patches.

Dispersal of the Allegheny woodrat is essential to understand, but difficult to study—For species functioning on a metapopulation level, it is dispersal that holds the population network together. Understanding the details of dispersal is vital for meaningful population prediction (Price et al. 1994, Lindenmayer and Lacy 2002). Although we have gained some understanding of individual behaviors at the subpopulation level, live-trapping and radiotelemetry studies of the Allegheny woodrat typically have focused at any one time on inhabitants of an individual rock patch. At this level, it is possible to detect immigrants and disappearances, but not to determine both origin and destination of dispersing woodrats. As a result, our picture of dispersal behavior and success in this species is still based on largely anecdotal and circumstantial evidence. Radiotelemetry studies conducted simultaneously at a group of outcrops, of the type being pioneered in West Virginia (Wood Chapter 3), will be necessary to get a better understanding of this phenomenon.

Diet of the Allegheny woodrat is varied, but not indiscriminate; some food resources are essential—The wide variety of items found in Allegheny woodrat food caches and described nonquantitatively in early studies made *N. magister* appear a dietary generalist. However, more quantitative recent studies have shown strong reliance on fungi in all seasons and on hard mast through the winter (Castleberry and Castleberry Chapter 6). We know virtually nothing about the ecological interactions between woodrats and fungi, although studies with other small mammal species are beginning to show complex relationships (Orrock and Pagels 2002, Vernes et al. 2004). Whether woodrats selectively harvest species of epigeous or hypogeous fungi, depend on fungi associated with particular forest plants, or choose fungus-contaminated foods for their palatability or digestibility are all open questions.

Dependence on hard mast is somewhat better understood from studies that show hard mast, especially acorns, is favored in the diet of the Allegheny woodrat (Castleberry and Castleberry Chapter 6). Hard mast represents a high-quality, nonperishable food which is a major component of caches. The ability to gather and defend a hard-mast cache is likely to affect over-winter survival and reproductive success the following spring, although this effect has yet to be clearly documented. Radiotelemetry studies demonstrate that the foraging ranges of both male and female Allegheny woodrats expand in summer during the cache-provisioning season, and contract in fall and winter when the food cache is being consumed. The observation that foraging activity and home range size increase in years of scarce hard mast, underlines the importance of hard mast as a dietary component and winter energy source. Soft mast and foliage also play a role in Allegheny woodrat diets, and their importance is not certain. Clearly, we need to understand how the diet of *N. magister* in declining populations differs from that in healthy populations, and whether food resource enhancement would provide a significant benefit.

Response of Allegheny woodrats to anthropogenic disturbance is complex and probably depends on spatial and temporal scale—Readers of this volume may be puzzled by apparently contradictory statements about the effects of human disturbance—factors such as habitat fragmentation, logging, wind farm activities, and residential development. LoGiudice (Chapter 2) cites habitat reduction and food resource depletion as contributory to woodrat decline, and Hassinger et al. (Chapter 8) present evidence that woodrat populations are most persistent if rock outcrops are embedded within an undisturbed forest matrix, whereas Castleberry (Chapter 4) cites studies showing that clearcutting and other logging practices have little effect on Allegheny woodrat foraging and movements. One of our study sites in central Pennsylvania was a popular hawk watch-spot for many years, where woodrats willingly took handouts from birders (J. Wright, pers. obs.). In western Pennsylvania, woodrats have successfully colonized abandoned quarry and rock dump sites (J. Peles pers. obs.).

The question of whether disturbance is beneficial or harmful for Allegheny woodrats is difficult to answer. For some types of disturbance there may be threshold effects, such as tolerance for varying canopy cover beyond a minimum necessary threshold argued by Castleberry (Chapter 4). Also, it may be that adaptability in the short run is distinct from susceptibility in the long run. When habitat is altered, a species may still live and forage there, but risk of predation and foraging success may change, inducing a barely perceptible decrease in survivorship and reproductive success that becomes catastrophic as population size declines. In particular, disturbances that make successful dispersal difficult may be breaking down, imperceptibly but inexorably, metapopulation networks throughout the Allegheny woodrat range. The indirect effects of human-caused disturbances over the past century—such as chestnut blight (*Cryphonectria parasitica*), gypsy moth (*Lymantria dispar*) invasion, and possible spread of raccoon roundworm through deliberate translocation of raccoons (*Procyon lotor*) by hunt clubs—may not be manifested for decades, or may be devastating only in combination with other factors. As it is clearly not feasible to limit all human activity in the interest of woodrat conservation, we need to tease out the relative impacts of different stressors under different conditions in order to take most effective conservation action.

We lack population models to predict trajectories and test the effects of different conservation scenarios for the Allegheny woodrat—We can now say definitely that some Allegheny woodrat populations are declining while others are fluctuating at near critical levels (Wood Chapter 3, Mengak et al. Chapter 7). What we cannot say with confidence is what will occur in the future. An essential tool in conservation planning is the development of realistic, useful models to analyze what threatened populations will do in the future, with and without different sorts of intervention. The process is known as population viability analysis (PVA; Akçakaya and Sjögren-Gulve 2000). So far, however, there has not been a single published population model for *N magister*. Generic metapopulation modeling programs for PVA are readily available and, with suitable caveats, can be used for the Allegheny woodrat (Brook et al. 2000, Lindenmayer et al. 2003). Two appropriate types of metapopulation models take different approaches. Incidence function or patch-occupancy models use snapshots of presence–absence at clusters of habitat sites to predict future patterns of occupancy. In a study of pikas, Moilanen et al. (1998) used such a model to show how a seemingly unimportant set of stepping-stone habitat islands determined the fate of a large complex of subpopulations. By contrast, individual-based models use more traditional demographic inputs in a spatially explicit context to project long-term trends. Pergams et al. (2000) used such a model to envision a population reintroduction of an island endemic deer mouse (*Peromyscus maniculatus anacapae*).

Recommendations

Design investigations to delineate metapopulation areas for the Allegheny woodrat —It is now becoming clear that the metapopulation, not the inhabitants of an individual rock patch or cave, should be the focus of future Allegheny woodrat population studies and conservation measures. Subpopulations are naturally impermanent; to assess current status and to measure future trajectories, we must work at the metapopulation level. This is a daunting challenge, not only because the field work will be more extensive, but because simply defining a metapopulation will require more knowledge of habitat requirements, genetic connectedness, and geographic and ecological features that favor or block dispersal from patch to patch. Field techniques for live-trapping and radiotelemetry studies of *N. magister* are already well standardized, and some useful nontrapping surveys are being developed (Mengak et al. Chapter 7). To work on a larger metapopulation scale, GIS techniques will be increasingly important for data projection, analysis, and coordination among research groups. Of particular interest is determining the genetic structure of metapopulations, an issue discussed in the next section.

Conduct studies to determine how behaviors and spacing at a microscale influence population success of Allegheny woodrats—Current studies suggest that even if den sites at a rock outcrop are numerous, access to high-quality dens may be limiting if context conditions such as safe foraging and food resources have deteriorated. Future studies of den selection should be focused on understanding factors influencing den preference, and the influence of den choice on over-winter survival and subsequent reproductive success. In addition, it will be important to use radiotelemetry to assess the fate of subadult and adult females at rock outcrops where den sites are limiting.

Focus research efforts on determining the timing, nature, and causes of dispersal in the Allegheny woodrat—If natural recolonization by dispersal was historically the mechanism maintaining Allegheny woodrat metapopulations, we need to understand why it is failing now and how it can be fostered or replaced. Larger-scale and year-round radiotelemetry investigations are needed to advance our understanding of this phenomenon beyond the anecdotal stage. Future radiotelemetry studies need to include juvenile Allegheny woodrats and to track all the individuals in a complex of rock patches to detect the timing, source, and destination of intersite movements.

Conduct studies to test hypotheses about the relationship between hard mast production and population dynamics of Allegheny woodrats—The "food decline" hypothesis to explain Allegheny woodrat population failures (LoGiudice Chapter 2) is logical but only loosely supported by data. Controlled long-term manipulation studies of nonsupplemented versus food-supplemented sites would provide a test of the hypothesis that current Allegheny woodrat populations suffer from inadequate food supply, and should help distinguish between impacts of food supply on over-winter survival versus subsequent reproduction. To complement field studies, laboratory investigations examining energetics of pregnancy, lactation and growth, and feeding experiments to determine diet choice, will give a better basis for estimating food resource requirements in the field.

Determine the long-term effects of disturbances, such as timber harvesting, on population dynamics of Allegheny woodrats—Results of short-term studies suggest that timber harvesting does not impact the activity patterns of woodrats (Castleberry Chapter 4). However, we know little about how this kind of perturbation translates into effects that might be manifested over longer periods of time. For example, increased risk of predation while foraging in altered habitats or decreased energy yield due to reduced mast availability might affect processes such as survival and reproduction. Consequently, future investigations should be designed to examine both short-term (i.e., movement patterns) and long-term (i.e., survival and reproduction) over multiple seasons and years.

Use population modeling for PVA and test the effects of different conservation management programs for the Allegheny woodrat—Spatially explicit models that incorporate interactions between subpopulations across a landscape can allow biologists and conservation planners to explore population dynamics and the impact of different interventions for metapopulation species in peril (Lindenmayer et al. 2003). Much of the information to parameterize such models for the Allegheny woodrat exists. Long-term and intensive studies conducted over the past 20 years (Wood Chapter 3, Peles and Wright Chapter 5, Mengak et al. Chapter 7) have yielded reams of data on age-specific birth and death rates, and population size and composition. Data on dispersal have been more elusive, but patterns are beginning to emerge (Wood Chapter 3). Landscape-level data on habitat site size and spatial arrangement have not been compiled in a consistent or systematic manner, but the need for such information and means to collect it are clear. We can make better use of existing data to make realistic population projections, but some difficult issues are involved. The issues include (1) assessing the validity of combining data from different sites to increase sample size for parameter estimation; (2) determining whether demographic rates and population sizes from "healthy" parts of the Allegheny woodrat range are demonstrably different from those in areas of decline; (3) determining the most biologically relevant measures of habitat patch size and quality; and (4) establishing criteria to distinguish "active" from "inactive" sites for patch-occupancy models when surveys have been conducted gradually over a period of years instead of simultaneously. Population modeling has definite limitations and its accuracy must always be checked with field data (Brook et al. 2000, Lindenmayer et al. 2003), but even clearly inaccurate results have led to useful insights about important variables that need greater study (Lindenmayer and Lacy 2002). This is a tool that should not be overlooked in conservation planning for the Allegheny woodrat.

Genetics of the Allegheny Woodrat

Lessons Learned

Genetic variation is greatly reduced in some small, geographically isolated populations of the Allegheny woodrat—Genetic variation in three populations of the

Allegheny woodrat from Indiana (Smyser and Rhodes Chapter 9) and another from New Jersey (Castleberry et al. 2002) is greatly reduced compared to average measures of heterozygosity and percent polymorphic loci in this species (Castleberry et al. 2002). The populations in New Jersey and Indiana are small and geographically isolated because of a decline in abundance and distribution of woodrats in those states. Remnant populations in Ohio and at least some parts of Pennsylvania share these characteristics, and also are likely to exhibit reduced genetic variation. Consequently, these populations, as well as others, may be at increased risk of extinction for several reasons. Reduced genetic variation resulting from genetic drift might limit the ability of a population to adapt to future environmental changes (Lande 1988). Reduced genetic variation may also reflect inbreeding—which has numerous effects on the fitness of individuals (Lande 1988)—within these isolated populations. Lastly, heterozygosity of individuals has been positively correlated with the expression of numerous fitness-related characteristics (Mitton and Grant 1984), and loss of heterozygosity presumably results in decreased fitness of individuals. These considerations increase the urgency for the initiation of well-planned reintroduction programs aimed at increasing the genetic diversity of remnant populations (Smyser and Rhodes Chapter 9, Serfass Chapter 10).

Gene flow in Allegheny woodrats is limited even at a small geographic scale—Gene flow is limited in Allegheny woodrats, even among individual rock outcrops, and the rate of gene flow decreases with distance (Wood Chapter 3, Smyser and Rhodes Chapter 9). This limited gene flow has been documented within the core of the species' range (i.e., West Virginia) where populations remain relatively stable. Restricted gene flow in these populations likely reflects habitat specificity and limited dispersal ability—characteristics inherent to the biology of the Allegheny woodrat. These characteristics likely increase the susceptibility of extant populations to extinction via genetic drift and inbreeding (Lande 1988, Belovsky et al. 1994) where: (1) local extinctions increase the distance between occupied outcrops; or (2) alteration of the landscape due to human activities further decreases the likelihood of movement between outcrops. Furthermore, this natural tendency toward limited gene flow and small effective population size reduces the probability of successfully re-establishing viable populations via reintroduction (Shafer 1981).

Recommendations

Use molecular genetics to define the boundaries of Allegheny woodrat metapopulations—Metapopulations have been viewed as appropriate management units for many species (Harrison 1994, Gilpin 1996). Wells and Richmond (1995) have argued that the first step in identifying a metapopulation is recognition of spatial disjunction between groups of populations. In fact, this is the approach that has been used for identifying woodrat metapopulations as part of the management plan for the state of Pennsylvania (Hassinger et al. Chapter 8). However, significant genetic differentiation has been demonstrated even within geographic clusters of rock outcrops (Smyser and Rhodes Chapter 9). These findings suggest that spatial arrangement alone may not be a sufficient criterion for defining a metapopulation in this species. That is, an area appearing to be spatially contiguous may actually be composed of multiple metapopulations because of the limited dispersal ability of the Allegheny woodrat. Genetic studies within presumed metapopulations are needed to define the boundaries of metapopulations in the Allegheny woodrat. In fact, we suggest that *N. magister* represents a model species for refining our understanding of the metapopulation concept.

Determine levels of genetic variation and patterns of genetic differentiation throughout the range of the Allegheny woodrat—The genetics of Allegheny woodrats have been studied within only a portion of the species' range (Castleberry et al. 2002, Smyser and Rhodes Chapter 9). However, conservation efforts for the Allegheny woodrat would greatly benefit from a genetic survey throughout the entire range of the species for two reasons. First, knowledge of rangewide patterns of genetic variation is necessary for making decisions regarding the source of animals for reintroductions (Serfass Chapter 10). Secondly, a possibility exists that woodrat populations in some regions may be following different evolutionary trajectories than those in other areas. Knowledge regarding patterns of genetic differentiation across the species' range would allow the identification of evolutionarily significant units that are useful for focusing conservation efforts below the species or subspecies level (see Fraser and Bernatchez 2001).

Conduct a comprehensive genetic survey to evaluate the taxonomic status of the Allegheny woodrat throughout its range—As documented in this book (Wright Chapter 1), conservation of a species may be hindered by a lack of understanding concerning its taxonomy. Because a comprehensive, rangewide genetic survey of *N. magister* is lacking, our understanding of the taxonomy of this species remains incomplete. As summarized by Wright (Chapter 1), genetic analyses are useful for distinguishing among subspecies or species along range borders (Ray 2000, Ray et al. 2002). We suggest that a genetic survey of *N. magister* throughout its range would benefit our understanding of taxonomy in two ways. First, these surveys would permit the evaluation of whether woodrat populations in some regions have become sufficiently differentiated to warrant subspecies status. Secondly, these surveys would permit evaluation of the extent of the species' range—especially in states such as North Carolina, Alabama, and Tennessee.

Conservation and Management Tools

Lessons Learned

Monitoring of the Allegheny woodrat should have been combined with formal hypothesis testing—Once *N. magister* was recognized as a declining species, biologists were faced with the monumental task of assessing current distributions and population numbers, while trying to understand the reasons for the decline of a species about which little was known concerning its biology. By the mid-1980s, several states had established programs for surveying potential woodrat habitat and monitoring populations (Mengak et al. Chapter 7). However, most efforts were concentrated on tracking the trajectory of the decline without attention to tests of possible correlative variables. For example, even crude assessments of factors such as the prevalence of *B. procyonis* or mast supply with respect to the presence and abundance of woodrats might have increased our understanding of the reasons for the decline and provided a basis for making conservation decisions.

Monitoring programs for the Allegheny woodrat would have benefited from communication among states—Monitoring of changes in the distribution and abundance of individual species is an essential aspect of conservation biology (Goldsmith 1991). As discussed by Mengak et al. (Chapter 7), management agencies in Indiana, Maryland, Pennsylvania, and Virginia established programs in the mid-1980s for surveying and monitoring woodrat populations in those states. The programs employed protocols that reflected the goals of each management agency and resulted in the collection of meaningful data and the development of beneficial approaches for the individual state. However, communication and coordination among states would have permitted standardization of at least some aspects of data collection (e.g., frequency of site visits, trapping effort, latrine counts). Standardization of data collection would have provided comparative data that could be used to evaluate temporal changes in abundance and distribution among states. In addition, standardized procedures would now be available for rapid implementation of monitoring programs by other states.

Previous reintroduction efforts for the Allegheny woodrat were conducted on a limited spatial and temporal scale—Serfass (Chapter 10) described the failures of previous reintroduction projects for the Allegheny woodrat in New Jersey, New York, Ohio, and Pennsylvania. Although two of these reintroductions involved more than one site, each of the sites represented individual rock outcrops that were relatively isolated. In addition, only one reintroduction effort involved the release of animals at more than one point in time. It is becoming increasingly apparent that a single release of Allegheny woodrats at the scale of the individual rock outcrop or habitat site is unlikely to result in the establishment of a viable population. The normal size of a population at an individual outcrop is relatively low and subject to stochastic extinction (Wood Chapter 3), behavioral interactions among individuals are complex (Peles and Wright Chapter 5), and expected mortality is high (Wood Chapter 3). Thus, it is not reasonable to expect that a one-time infusion of a few or even a large number of woodrats at an individual rock outcrop will establish a viable population. The failure of previous reintroductions illustrates the need to focus efforts at the metapopulation level (i.e., multiple rock outcrops linked by dispersal) over a time period (i.e., multiple years) that permits supplementation (Serfass Chapter 10).

Recommendations

Combine the best features of individual state programs to create a standardized process for monitoring populations of the Allegheny woodrat in each state—As described by Mengak et al. (Chapter 7), the long-term monitoring programs in Indiana, Maryland, Pennsylvania, and Virginia used somewhat varied approaches and emphasized different aspects of the monitoring process. However, each of these programs had many positive features. We suggest that the positive aspects of each program be used to create a standardized process for monitoring and surveying Allegheny woodrats in each state. While we recognize that each state would have to tailor the process to meet specific needs and goals, this standardized process would facilitate coordination and cooperation among states. Future monitoring of the Allegheny woodrat will be critical for recognizing changes in distribution and abundance in areas (e.g., Pennsylvania) where a decline is already in progress and assessing whether a decline is occurring in other states also. This latter point is especially important in states, such as Virginia (Mengak et al. Chapter 7) and West Virginia (Wood Chapter 3), where there is at least some evidence to suggest a decrease in the abundance or distribution of woodrats. Furthermore, Hassinger et al. (Chapter 8) have pointed out the importance of detecting declines in states within the southern portion of the species' range because this might serve as the impetus for federal listing of the Allegheny woodrat. Lastly, a standardized monitoring program will be essential for evaluating the success of management and reintroduction programs for woodrats, and could also be used as a model for other species (Goldsmith 1991, Thompson et al. 1998).

Use the model developed by the Pennsylvania Game Commission as the basis for a comprehensive management program for the Allegheny woodrat—To date, conservation and management efforts for the Allegheny woodrat within individual states have been limited to monitoring and assessment of abundance and distribution. In fact, the use of the term "management" for this species remains relatively vague. Here, we use this term to refer to efforts to conserve existing populations of woodrats, preserve existing woodrat habitat, to reverse or offset factors that are negatively influencing the species, and develop plans for reintroduction or supplementation. Hassinger et al. (Chapter 8) have presented a model program based on their experience with woodrats in Pennsylvania. This model complements the recent state Wildlife Action Plans in that it stresses the importance of habitat-based management. We suggest that the Pennsylvania management program should be used as a basis for establishing a comprehensive, multistate management program for the Allegheny woodrat. This program also should involve cooperation among those concerned with management of other resources, such as forestry and game species (Hassinger et al. Chapter 8).

Use adaptive resource management approaches in planning and research for the Allegheny woodrat—Producing management plans is important, but no plan is written in stone. In conservation biology, even the best plans are based in part on limited data and on assumptions and generalizations, so that "most policies are really questions masquerading as answers" (Gunderson 1999). Realizing the need for more research, Hassinger et al. (Chapter 8) advocate an adaptive resource management approach that includes data gathering with a periodic review of the management plan to allow appropriate adjustments. Similarly, PVA modeling profits from iterative tests against real data with adjustments of the model to make it more realistic and useful. The monitoring programs described in Chapter 7, and recommended ecological studies described in this chapter, would be excellent reality checks for predictive models and management plans and should be integral parts of them.

Develop a plan to establish priorities and procedures for an Allegheny woodrat reintroduction program—Chapters in this volume summarize the plight of the Allegheny woodrat—populations have been extirpated from some states or geographic regions (Wright Chapter 1, LoGiudice Chapter 2, Hassinger et al. Chapter 8), and others have declined in numbers (Mengak et al. Chapter 7). In addition, there is evidence that at least some populations have experienced a reduction in genetic variation (Smyser and Rhodes Chapter 9). Consequently, we suggest that any single- or multi-state management plan should include an evaluation of the potential for reintroduction or supplementation of woodrats. We recommend that any such program should be focused on the metapopulation level. Several woodrat reintroduction programs have already been unsuccessful (Serfass Chapter 10). Thus, careful consideration and planning, as outlined by Serfass (Chapter 10), must be given to any reintroduction project. It should also be noted that few reintroduction projects have been conducted with *r*-selected (short-lived, high-mortality) species. Planning reintroductions in a metapopulation context is even more daunting, and may take novel approaches such as repeated small-volume translocations (Lubow 1996). While the task is difficult, it also presents an opportunity to use the Allegheny woodrat as model species for developing a reintroduction protocol specifically for *r*-selected species.

References

Akçakaya H. R., and Sjögren-Gulve, P. 2000. Population viability analysis in conservation planning: an overview. Ecological Bulletins 48:9–21.

Beans, B. E. 1992. Without a trace: the puzzling demise of the Allegheny woodrat. Audubon 94:32–34.

Belovsky, G. E., Bissonette, J. A., Dueser, R. D., Edwards, T. C. Jr., Luecke, C. M., Ritchie, M. E., Slade, J. B., and Wagner, F. H. 1994. Management of small populations: concepts affecting the recovery of endangered species. Wildlife Society Bulletin 22:307–316.

Brook, B. W., O'Grady, J. J., Chapman, A. P., Burgman, M. A., Akçakaya H.R, and Frankham, R. 2000. Predictive accuracy of population viability analysis in conservation biology. Nature 404:385–386.

Castleberry, S. B., King, T. L., Wood, P. B., and Ford, W. M. 2002. Microsatellite DNA analysis of population structure in Allegheny woodrats (*Neotoma magister*). Journal of Mammalogy 83:1058–1070.

Ceballos, G., and Brown, J. H. 1995. Global patterns of mammalian diversity, endemism, and endangerment. Conservation Biology 9:559–568.

Franken, R. J., and Hik, D. S. 2004. Influence of habitat quality, patch size and connectivity on colonization and extinction dynamics of collared pikas *Ochotona collaris*. Journal of Animal Ecology 73:889–896.

Frase, B. A., and Hoffman, R. S. 1980. *Marmota flaviventris*. Mammalian Species 135:1–8.

Fraser, D. J., and Bernatchez, L. 2001. Adaptive evolutionary conservation: towards a unified concept for defining conservation units. Molecular Ecology 10:2741–2752.

Gerlach, G., and Hoeck, H. N. 2001. Islands on the plains: metapopulation dynamics and female biased dispersal in hyraxes (Hyracoidea) in the Serengeti National Park. Molecular Ecology 10:2307–2317.

Gilpin, M. 1996 Metapopulations and wildlife conservation: approaches to modeling spatial structure. *In* Metapopulations and Wildlife Conservation, ed. D. R. McCullough, pp. 11–27. Island Press, Washington, D.C.

Goldsmith, F. B. (ed.). 1991. Monitoring for Conservation and Ecology. Chapman and Hall, London, United Kingdom.

Gonzalez, A., Lawton, J. H., Gilber, F. S., Blackburn, T. M., and Evans-Freke, I. 1998. Metapopulation dynamics, abundance, and distribution in a microecosystem. Science 281:2045–2047.

Gunderson, L. 1999. Resilience, flexibility and adaptive management: antidotes for spurious certitude? Conserveration Ecology 3:7.

Hanski, I. 1996. Metapopulation ecology. *In* Population Dynamics in Ecological Space and Time, eds. O. E. Rhodes, Jr., R. K. Chesser and M. H. Smith, pp. 13–43. University of Chicago Press, Chicago, Illinois.

Harrison, S. 1994. Metapopulations and conservation. *In* Metapopulation Dynamics, eds. M. Gilpin and I. Hanski, pp. 111–130. Blackwell Publishing, Oxford, United Kingdom.

Hayes, J. P., and Harrison, R. G. 1992. Variation in mitochondrial DNA and the biogeographic history of woodrats (*Neotoma*) of the eastern United States. Systematic Biology 41:331–344.

Hayward, M. W., de Tores, P. J., Dillon, J. J., and Fox, B. J. 2006. Local population structure of a naturally occurring metapopulation of the quokka (*Setonix brachyurus* Macropodidae: Marsupialia). Biological Conservation 110:343–355.

Hicks, A. P. 1989. Whatever happened to the Allegheny woodrat? New York State Conservationist 43:34–38.

Hoogland, J. L. 1996. *Cynomys ludovicianus*. Mammalian Species 535:1–10.

Isaac, N. J. B., Mallet, J., and Mace, G. M. 2004. Taxonomic inflation: its influence on macroecology and conservation. Trends in Ecology and Evolution 19:464–469.

Kinsey, K. P. 1976. Social behaviour in confined populations of the Allegheny woodrat, *Neotoma floridana magister*. Animal Behaviour 24:181–187.

Kinsey, K. P. 1977. Agonistic behavior and social organization in a reproductive population of Allegheny woodrats, *Neotoma floridana magister*. Journal of Mammalogy 58:417–419.

Krohne, D. T. 1997. Dynamics of metapopulations of small mammals. Journal of Mammalogy 78:1014–1028.

Lande, R. 1988. Genetics and demography in biological conservation. Science 241:1455–1460.

Lawes, J. J., Mealin, P. E., and Piper, S. E. 2000. Patch occupancy and potential metapopulation dynamics of three forest mammals in fragmented Afromontane forest in South Africa. Conservation Biology 14:1088–1098.

Lindenmayer, D. B., and Lacy, R. C. 2002. Small mammals, habitat patches and PVA models: a field test of model predictive ability. Biological Conservation 103:247–265.

Lindenmayer, D. B., Possingham, H. P., Lacy, R. C., McCarthy, M. A., and Pope, M. L. 2003. How accurate are population models? Lessons from landscape-scale tests in a fragmented system. Ecology Letters 6:41–47.

LoGiudice, K. 2006. Toward a synthetic view of extinction: a history lesson from a North American rodent. BioScience 56:687–693.

Lubow, B. C. 1996. Optimal translocation strategies for enhancing metapopulation viability. Ecological Applications 6:1268–1280.

Marsh, L. K., and Wagner, E. G. (eds.). 2003. Primates in Fragments: Ecology and Conservation. Plenum Publishers, New York.

Martin, J. M., Heske, E. J., and Hoffman, J. E. 2003. Franklin's ground squirrel (*Spermophilus franklinii*) in Illinois: a declining prairie mammal? American Midland Naturalist 150: 130–138.

Mitton, J. B., and Grant, M. C. 1984. Associations among protein heterozygosity, growth rate, and developmental homeostasis. Annual Review of Ecology and Systematics 15:479–499.

Moilanen, A., Smith, A. T., and Hanski, I. 1998. Long-term dynamics in a metapopulation of the American pika. The American Naturalist 152:530–542.

Orrock, J. L., and Pagels, J. F. 2002. Fungus consumption by the southern red-backed vole (*Clethrionomys gapperi*) in the southern Appalachians. American Midland Naturalist 147:413–418.

Ozgul, A., Armitage, K. B., Blumstein, D. T., Van Vuren, D. H., and Oli, M. K. 2006. Effects of patch quality and network structure on patch occupancy dynamics of a yellow-bellied marmot metapopulation. Journal of Animal Ecology 75:191–202.

Pergams, O. R. W., Lacy, R. C., and Ashley, M. V. 2000. Conservation and management of Anacapa Island deer mice. Conservation Biology 14:819–832.

Pimm, S. L. 1996. Lessons from a kill. Biodiversity and Conservation 5:1059–1067.

Price, M. V., Kelly, P. A., and Goldingay, R. L. 1994. Distances moved by Stephens' kangaroo rat (*Dipodomys stephensi* Merriam) and implications for conservation. Journal of Mammalogy 75:929–939.

Ray, D. K. 2000. Phylogenetics and evolution of woodrats (Genus *Neotoma*) in the southern Appalachian Mountains. M.S. Thesis, University of North Carolina at Wilmington, Wilmington, North Carolina.

Ray, D. K., Freshwater, D. W., and Webster, W. D. 2002. Multiplex PCR identification of eastern and Allegheny woodrats. Journal of Wildlife Management. 66:86–90.

Roach, J. L., Stapp, P., Van Horne, B., and Antolin, M. F. 2001. Genetic structure of a metapopulation of black-tailed prairie dogs. Journal of Mammalogy 82:646–959.

Shafer, M. L. 1981. Minimum population sizes for species conservation. Bioscience 31:131–134.

Sweanor, L. L., Logan, K. A., and Hornocker, M. G. 2000. Cougar dispersal patterns, metapopulation dynamics, and conservation. Conservation Biology 14:798–808.

Telfer, S., Holt, A., Donaldson, R., and Lambin, X. 2001. Metapopulation processes and persistence in remnant water vole populations. Oikos 95:31–42.

Thompson W. L., White, G. C., and Gowan, C. 1998. Monitoring Vertebrate Populations. Academic Press, Burlington, Massachusetts.

Vernes, K., Blois, S., and Bärlocher, F. 2004. Seasonal and yearly changes in consumption of hypgeous fungi by northern flying squirrels and red squirrels in old-growth forest, New Brunswick. Canadian Journal of Zoology 82:110–117.

Wells, J. V., and Richmond, M. E. 1995. Populations, metapopulations, and species populations: what are they and who should care? Wildlife Society Bulletin 23:458–462.

Index

A

Acer spp., *see* Maple (*Acer spp.*)
Acorns
  competition for, 28, 141
  in diet of Allegheny woodrat, 94, 95, 214
  reduced availability as factor in decline, 26, 119, 120
  replacement of chestnuts in diet, 34
  and reproduction, 143
  as supplemental food source, 26, 27, 202
Activity center, *see* Population units, for woodrat research and management
Adaptive resource management, 137, 221
Agonistic behavior, *see* Behavior, agonistic
Alabama, history and status of Allegheny woodrat in, 15–17
Amelanchier canadensis, *see* Juneberry (*Amelanchier canadensis*)
American Chestnut Cooperators Foundation, 144
American Chestnut Foundation, 144
Antihelmintics, 141, 142, 211
Arthropods, as food item, 96

B

Baird, S. F., 4–8
Basswood, American (*Tilia americana*), as food item, 94, 101
Bat gates, as protection for cave habitat, 142
Baylisascaris procyonis, *see* Raccoon roundworm (*Baylisascaris procyonis*)
Bear, black (*Ursus americanus*), 23, 33, 35, 134, 141, 143, 146, 147
Beech, American (*Fagus grandifolia*), as food item, 94, 101
Behavior
  agonistic, 76, 83, 185, 186, 211
  caching, 96, 201
  and conservation, 7, 89, 90, 212
  as consideration for reintroduction plan, 184–186
  as factor in decline, 75, 201
  foraging and home range, 63
  and risk of exposure to raccoon roundworm, 31, 201
Betula spp., *see* Birch (*Betula* spp.)
Birch (*Betula* spp.)
  as food item, 94, 101
  replacement of oak species, 28, 98, 143
Blackberry (*Rubus* spp.), as food item, 94, 98
Blueberry (*Vaccinium* spp.), as food item, 94, 101, 103
Bubo virginianus, *see* Owl, great horned (*Bubo virginianus*)

C

Cache(s), 13, 85, 96, 97, 111, 135, 141, 184, 201, 202, 214
CAPTURE (population model), 126
Carya spp., *see* Hickory (*Carya* spp.)
Castanea dentata, *see* Chestnut, American (*Castanea dentata*)
Cats, feral, as predators, 196–198
Cave rat, 6, 8
Cave(s), as habitat, 134, 142, 146
Central-place foraging, 70
Chestnut, American (*Castanea dentata*)
  destruction as factor in decline, 24, 27, 34, 97, 103, 133, 143
  as food item, 95, 101
  restoration of, 144
  as stable food source, 27, 28, 95, 98, 117, 143, 144
Chestnut blight (*Cryphonectria parasitica*), 24, 27, 28, 97, 117, 143, 165, 215
Connecticut
  gypsy moth in, 35

Connecticut (*Cont.*)
history and status of Allegheny woodrat in, 13, 14, 17
loss of American chestnut in, 34
raccoon roundworm in, 31
Conservation management area, *see* Population units, for woodrat research and management
Cotton rat, hispid (*Sigmodon hispidus*), 27
Crotalus horridus, *see* Rattlesnake, timber (*Crotalus horridus*)
Cryphonectria parasitica, *see* Chestnut blight (*Cryphonectria parasitica*)
Cryptic species, and conservation, 17

D
Decline
of eastern woodrat, in Illinois, 198, 199
of insular woodrat species, 196
of Key Largo woodrat, 180, 198
of mainland woodrat species, reasons for, 201
of riparian woodrat, 197
Decline, of the Allegheny woodrat
critical stressors related to, 133, 137
detecting/documenting, 3, 12, 13, 18, 172, 210, 211
history of, 14–17, 24, 25
hypotheses to explain, 26–33, 97, 98, 103, 140, 159
hypothetical chronology, 33–36
lessons learned from, 210, 211
recent, 56, 114, 117, 123, 155, 159
recommendations based on, 211, 212
as result of interacting factors, 24, 133, 211
Deer, white-tailed (*Odocoileus virginianus*), 23, 24, 28, 33–35, 98, 133, 141, 143, 144, 169
Deforestation, and decline of Allegheny woodrat, 33, 34
Den
construction, 76
fidelity, 82
as limiting resource, 81, 89, 216
preference, 81, 216
quality, 81, 87, 213
seasonal shifting, 84, 184
Den location
central versus peripheral, 80, 87, 213
as determined by radiotelemetry, 79
males versus females, 80, 213
spacing of, 83, 84, 87
Diet, of the Allegheny woodrat, 99–102
diversity in, 103
generalist, 142, 143, 214
green vegetation in, 96
hard mast in, 28, 94, 214
historical importance of chestnuts in, 27, 28
methods of analysis, 93, 94
and reproduction, 28
seasonal, 94–96
seeds and fruits in, 28
soft mast in, 94, 95, 214
Dispersal, of Allegheny woodrat
adults, 49, 58
barriers to, 47, 58, 136, 158
and breeding, 47
distance, 46, 47, 135, 136, 142, 145
effects of deforestation/habitat fragmentation on, 29, 33, 34, 35, 133, 139
following reintroduction, 162, 185, 203, 220
and gene flow, 164
juveniles, 48, 58
in metapopulations, 12, 36, 152, 157, 158, 214–216
natal, 47, 59
and selection of reintroduction site, 175, 176, 180
Dispersal corridors, 135, 139, 142, 145, 152, 158, 164

E
Elaphe obsoleta, *see* Snake, black rat (*Elaphe obsolete*)
Endangered species, of Neotoma, 195–198
Extinction vortex, 161
Extirpation
of Allegheny woodrat populations in Indiana, 159, 161, 164
and distance to nearest occupied site, 24
of eastern woodrat in southern Illinois, 203
at habitat sites, 113, 118, 123
of a metapopulation, 113, 119
rate, of woodrat populations in Maryland, 120
vulnerability of small populations to, 36

F
Fagus grandifolia, *see* Beech, American (*Fagus grandifolia*)
Feces, 141, 142, 201
Fern(s), 94, 141, 143
Fire suppression, 98, 143
Food cache, *see* Cache(s)
Food decline hypothesis, 26, 27, 216
Food, *see* Diet, of the Allegheny woodrat

Foraging
buffer zone, 135, 136, 142, 144, 176
changes in response to disturbance, 68, 69, 211
changes in response to mast availability, 69, 214
distance and rate, 65, 143, 214, 215, 217
Forest
canopy cover and use, 68–70
clearcutting, 57, 67, 140, 141
Fossil(s), 4–8, 13, 25, 159
Fox, gray (*Urocyon cinereoargentus*), 141
Fox, red (*Vulpes vulpes*), 141
Fungi
as food item, 94, 95, 99, 214
nutritive value of, 95

G

Gene flow
by dispersal, 164
and disturbance, 70, 164
evidence for, 50
historic patterns of, 178
limited, among subpopulations, 157, 218
by translocation, 161
Genetic differentiation, 50, 157, 161, 212, 218, 219
Genetic diversity
in Allegheny woodrat, 157–163
in eastern woodrat from Illinois, 199, 201
enhancement in small populations, 145
maintenance in metapopulations, 155, 156
and population persistence, 164
in reintroduced populations, 187, 202
and reintroduction plan, 178–180, 218
Genetic drift, 153–155, 157, 158, 163, 164, 201
Georgia
history and status of Allegheny woodrat, 15
raccoon roundworm in, 32
GIS (Geographic Information Sysems)
as basis for mapping museum specimens, 4
utility of, 18, 114, 115, 174, 175, 177, 203, 216
Grape (*Vitis* spp.), 94, 98, 103, 144
Gum, black (*Nyssa sylvatica*), as food item, 94, 98
Gypsy moth (*Lymantria dispar*), 26, 28, 34, 35, 97, 111, 120, 133, 134, 143, 165, 215

H

Habitat
of the Allegheny woodrat, 133, 135, 210
destruction/loss, 24, 29, 34, 69, 134, 139, 140, 164, 196–198
extent in Indiana, 121
extent in Pennsylvania, 113
fragmentation, 29, 34, 36, 120, 139–141, 154, 164, 198, 201, 213, 214
restoration, 197, 203
selection, 66–67
Habitat fragmentation/destruction hypothesis, 29, 30
Habitat site, *see* Population units, for woodrat research and management
Hantavirus, in woodrats, 183
Hard-release technique, for reintroduction, 185
Hemlock, eastern (*Tsuga canadensis*), as food item, 94, 98
Hickory (*Carya* spp.)
as food item, 95, 101
nutritive value of, 94
Historical-ecological approach, 25, 37
Historical-ecological hypothesis, 33
Hoarders, conservation implications, 98, 201
Holly (*Ilex* spp.), as food item, 94
Home range, size, 46, 63, 143, 213, 214
Human disturbance, and decline of the Allegheny woodrat, 34, 214

I

Ilex spp., *see* Holly (*Ilex* spp.), as food item
Illinois
decline of eastern woodrat in, 198
genetic diversity of eastern woodrat in, 199
reintroduction of eastern woodrat in, 199–201
Inbreeding depression, 153, 154, 159, 164, 201
Indiana
decline of Allegheny woodrat in, 13, 17, 123, 159, 160
genetic diversity in, 160–163
history and status of Allegheny woodrat, 13, 17, 121
loss of American chestnut in, 34
monitoring of Allegheny woodrat in, 121–124, 159, 160
raccoon roundworm in, 30–32
Introgression, 11
Inventorying, of abundance and distribution, 109, 113, 116, 117, 122, 125
Island endemic species, of Neotoma, 195–197
Isoflurane, 184
IUCN
and conservation status, 12
criteria for endangered species, 138
guidelines for reintroduction projects, 171–187

J

Juglans nigra, *see* Walnut, black (*Juglans nigra*), as food item
Juneberry (*Amelanchier canadensis*), as food item, 94, 101

K

Karyotype, 11
Kentucky, history and current status of Allegheny woodrat in, 16, 67
Ketamine hydrochloride, 184
K-selected species, 170, 171, 188

L

Latrine(s), 13, 30, 31, 36, 111, 116, 119, 135, 220
Leopard, clouded (*Neofelis diardi*), 3
Lethal equivalents, 155
Life span, maximum, 51, 117
Lincoln-Peterson population estimate, 117, 126
Live-trapping, 14, 54, 78–79, 111, 115, 121, 125, 159, 186, 200, 202, 214, 216
Long-term monitoring (LTM), 110, 111, 113, 117, 122, 125, 186, 202, 221
Lymantria dispar, *see* Gypsy moth (*Lymantria dispar*)
Lyme disease, in woodrats, 183

M

Maple (*Acer* spp.)
  as food item, 94, 101, 103
  replacement of oak species, 28, 98, 143, 144
Maryland
  decline of Allegheny woodrat in, 26, 117
  history and current status of Allegheny woodrat in, 14, 15, 17, 115
  loss of American chestnut in, 34
  monitoring of Allegheny woodrat in, 115–121, 220, 221
  raccoon roundworm in, 30
Massachusetts, history and status of Allegheny woodrat in, 13, 14
Mast
  availability, 56, 143, 211
  competition for, 33–36, 143
  effects on reproduction, 27, 28, 143
  influence of gypsy moth on availability, 35
  oak, *see* Acorns
  and population size, 56, 97, 120
  stability of oak versus American chestnut, 28, 98
Maxillovomerine notch, as diagnostic character, 11
Meleagris gallopavo, *see* Turkey (*Meleagris gallopavo*)
Mephitis mephitis, *see* Skunk, striped (*Mephitis mephitis*)
Metapopulation
  application to conservation, 109, 137, 164, 165, 216
  definition of, 12, 45, 110, 133
  detecting decline in, 12, 13
  dispersal in, 35, 36, 47, 155, 214
  extirpation of, 114, 118, 119, 145
  gene flow in, 164
  genetic differentiation within, 157, 158
  identification of, 218, 219
  maintaining genetic diversity in, 156, 160
  management of, 136, 137, 147
  recolonization of, 118, 119
  and reintroduction plan, 176, 177, 180, 188, 220, 222
  size and viability, 152
  structure of Allegheny woodrat, 45, 155, 163
Metapopulation area, *see* Population units, for woodrat research and management
Michella repens, *see* Partridgeberry (*Michella repens*), as food item
Midden(s), 143, 181
Mitochondrial DNA
  and genetic diversity, 156
  and taxonomy, 11, 15
Monitoring
  to detect future declines, 212
  of eastern woodrat, following reintroduction, 200, 202
  following reintroduction, 145, 185–187
Monitoring, of Allegheny woodrat
  in Indiana, 121–124, 159, 160
  in Maryland, 115–121
  in Pennsylvania, 111–115, 138
  in Virginia, 124–129
Monitoring programs
  defined, 109
  recommendations for, 129–131, 137, 148, 220–222
Morphological characters, and taxonomy, 5, 11
Mouse, white-footed (*Peromyscus leucopus*), 28, 35, 154
Museum collections, type specimens, 7
Museum specimens, 4–9, 18
Mustela frenata, *see* Weasel, long-tailed (*Mustela frenata*)

N

Nelsonia spp., *see* Woodrat, diminutive (*Nelsonia* spp.)
Neofelis diardi, *see* Leopard, clouded (*Neofelis diardi*)
Neotoma
  albigula, *see* Woodrat, white-throated (*Neotoma albigula*)
  albigula varia, *see* Woodrat, Turner Island, (*Neotoma albigula varia*)
  cinerea, *see* Woodrat, bushy-tailed (*Neotoma cinerea*)
  floridana illinoensis, 193, 198–201
  floridana magister (synonym for N. magister), 4
  floridana, *see* Woodrat, eastern (*Neotoma floridana*)
  floridana smalli, *see* Woodrat, Key Largo (*Neotoma floridana smalli*)
  fuscipes riparia, *see* Woodrat, riparian (*Neotoma fuscipes riparia*)
  fuscipes, *see* Woodrat, dusky-footed (*Neotoma fuscipes*)
  lepida abbreviata, 197
  lepida latirostra, 197
  lepida, *see* Woodrat, desert (*Neotoma lepida*)
  macrotis, *see* Woodrat, big-eared (*Neotoma macrotis*)
  magister, first description, 4
  micropus, *see* Woodrat, southern plains (*Neotoma micropus*)
  pennsylvanica (synonym for N. magister), 4, 5, 6–8, 11
Nest(s), of Allegheny woodrat, 13, 76
New Jersey
  gypsy moth in, 35
  history and status of Allegheny woodrat in, 14
  population size and mast production in, 26
  raccoon rabies in, 32
  raccoon roundworm in, 30, 31, 36
  reintroduction of Allegheny woodrat in, 30, 31, 170
New York
  gypsy moth in, 35
  history and status of Allegheny woodrat in, 13, 24
  loss of American chestnut in, 34
  raccoon roundworm in, 30, 35, 36
  reintroduction of Allegheny woodrat in, 30, 31, 145, 170
North Carolina, history and status of Allegheny woodrat in, 15, 17
Nuclear DNA, and genetic diversity, 156, 157, 199, 201
Nyssa sylvatica, *see* Gum, black (*Nyssa sylvatica*), as food item

O

Oak (*Quercus* spp.), 23, 25, 26, 28, 34, 94, 119, 120, 175, 197
Odocoileus virginanus, deer, white-tailed, 23, 102, 133, 169
Ohio
  history and status of Allegheny woodrat in, 15, 17
  loss of American chestnut in, 34
  reintroduction of Allegheny woodrat in, 170, 184, 185, 220
Optimal foraging, 97
Owl, great horned (*Bubo virginianus*), 29–35, 67, 141, 175

P

Parasite hypothesis, 30–32
Partners in Flight, 148
Partridgeberry (*Michella repens*), as food item, 94, 100
Pennsylvania
  decline of Allegheny woodrat in, 114
  den use behavior in, 76–85
  food habits in, 94
  gypsy moth in, 26, 35
  history and status of Allegheny woodrat in, 13, 14, 17, 24, 111–115
  loss of American chestnut in, 28, 34
  monitoring of Allegheny woodrat in, 111, 134, 138, 139
  raccoon roundworm in, 30
  reintroduction of Allegheny woodrat in, 145, 171, 220
  survival and life expectancy of Allegheny woodrat in, 179
Peromyscus leucopus, *see* Mouse, white-footed (*Peromyscus leucopus*)
Pine, white (*Pinus strobus*), as food item, 94, 98
Pinus strobus, *see* Pine, white (*Pinus strobus*), as food item
Population
  classification system, 134–136
  decline, *see* Decline
  fluctuation, 55, 113, 126, 127, 186
Population size
  of Allegheny woodrat, 86, 154, 157

Population size (*Cont.*)
and genetics, 153, 155, 157–159, 211, 218
in Indiana, 159
in Pennsylvania, 113
of riparian woodrat, 197
sustainable, 86
in Virginia, 126, 129
and vulnerability, 145
Population units, for woodrat research and management
activity center, 135
conservation management area, 112, 113
habitat site, 134–147, 152, 220
metapopulation area, 135–138, 152, 216
Postglacial expansion, 156
Predation
as mortality factor, 52, 175
risk of, 56, 70, 215, 217
Predators
of Allegheny woodrat, 133, 141
of island endemic woodrats, 196
of Key Largo woodrat, 198
PRESENCE (patch occupancy estimate), 128
Procyon lotor, *see* Raccoon (*Procyon lotor*)
PVA (Population Viability Analysis), 215, 217, 221

Q

Quercus spp., *see* Oak (*Quercus* spp.)

R

Raccoon (*Procyon lotor*), 16, 17, 30–36, 133, 141, 175, 184, 198, 201, 202, 215
Raccoon roundworm (*Baylisascaris procyonis*), 17, 30, 31, 34, 36, 52, 56, 97, 120, 133, 141, 142, 159, 175, 184, 198, 201–203, 211, 215
Radiotelemetry
detecting dispersal by, 77–79, 213
following a reintroduction, 171, 186, 200
home range determination by, 64
location of dens by, 77–79, 213
Range, of Allegheny woodrat
historic versus current, 12–17, 24–25
maps, 7–10
Rare species, determining distribution and abundance, 109
Rarity, and woodrats, 195
Rattlesnake, timber (*Crotalus horridus*), 138, 142, 146
Recolonization, 35, 36, 114, 117, 123, 145, 156, 164, 176, 179, 188, 212, 216
Reintroduction
of Allegheny woodrat in New Jersey, 26, 27, 170
of Allegheny woodrat in New York, 26, 145, 170
of Allegheny woodrat in Ohio, 170, 184, 185
of Allegheny woodrat in Pennsylvania, 145, 171
as an adaptive management approach, 145, 146
as conservation measure, 145
definition of, 169
of eastern woodrat in Illinois, 199–201
and genetic considerations, 162, 218, 219
guidelines for, 171–187, 202
limitations of previous attempts, 220
recommendations for, 221, 222
Reproduction
following reintroduction of eastern woodrat, 200
influence of food supplementation on, 27, 28, 143
influence of mast production on, 143, 217
Reproductive characteristics, of Allegheny woodrat, 53
Reproductive energetics, 27
Reproductive success, 54, 154, 186, 211, 214–216
Rocky habitat
home range size in relation to, 65
lack of decrease in, 29
and restriction of local distribution, 155
r-selected species, 170, 171, 179, 180, 187, 188, 222
Rubus spp., *see* Blackberry (*Rubus* spp.), as food item

S

Scat
raccoon, consumption by woodrats, 31
woodrat, *see* Latrine(s)
Secondary extinction, 24
Sigmodon hispidus, *see* Cotton rat, hispid (*Sigmodon hispidus*)
Site selection, for reintroduction, 152, 174–177
Skunk
eastern spotted (*Spilogale putoris*), 141
striped (*Mephitis mephitis*), 141
Snake, black rat (*Elaphe obsoleta*), 141
Soft-release technique, for reintroduction, 185, 200, 202
Source population, for reintroduction, 171, 173, 178, 179, 187

Spilogale putoris, *see* Skunk, eastern spotted (*Spilogale putoris*)
Stone, W., 5, 6
Supplemental feeding, 26, 27, 111, 113, 143, 144
Supplementations, 169, 179, 180, 187, 188
Surface rock communities, 134, 135, 141, 146–148
Surveying
  of Allegheny woodrat habitat, 137, 138, 220
  of Allegheny woodrat populations, 109, 113, 116, 117, 138, 220, 221
  for presence of raccoon roundworm, 152
  quantitative latrine and midden technieques, 116
  visual methods, 11, 110, 113, 114, 116
Survey techniques, 137, 138
Survival
  of adults, 52
  of Allegheny woodrat in Ohio following reintroduction, 184, 185
  of eastern woodrat in Illinois following reintroduction, 200
  importance to success of reintroduction program, 179, 180, 182, 183, 186
  and inbreeding, 154, 155
  influence of nutrition on, 27, 28
  influence of supplemental feeding on, 143
  of juveniles, 52
  over-winter, 52, 88, 119, 214, 216
  of *r*-selected species, 179

T
Taxonomic history, of the Allegheny woodrat, 4–12
Taxonomy
  and conservation, 18
  of Neotoma, 193–195
Tennessee, history and status of Allegheny woodrat in, 16, 17
Tilia americana, *see* Basswood, American (*Tilia americana*), as food item
Toilet area, *see* Latrine(s)
Translocation, *see* Reintroduction
Tsuga canadensis, *see* Hemlock, eastern (*Tsuga canadensis*), as food item
Turkey (*Meleagris gallopavo*), 24, 33–35, 141, 143

U
Urocyon cinereoargentus, *see* Fox, gray (*Urocyon cinereoargentus*)
Ursus americanus, *see* Bear, black (*Ursus americanus*)

V
Vaccinium spp., *see* Blueberry (*Vaccinium* spp.), as food item
Virginia
  history and status of Allegheny woodrat in, 15, 17
  home range size in, 64
  monitoring of Allegheny woodrat in, 124–129, 220, 221
Vitis spp., *see* Grape (*Vitis* spp.)
Vulpes vulpes, *see* Fox, red (*Vulpes vulpes*)

W
Walnut, black (*Juglans nigra*), as food item, 95, 101
Weasel, long-tailed (*Mustela frenata*), 141
Weather, 56, 154, 201, 203
West Virginia
  disturbance as predictor of presence, 67
  food habits in, 94
  history and status of Allegheny woodrat in, 14, 15, 17
  home range size in, 64
  population study methods in, 55, 214
  raccoon roundworm in, 31
Wildlife Action Plans (CWCS), 18, 147, 172, 221
Woodrat
  Anthony's (*Neotoma anthonyi*), 195, 196
  big-eared (*Neotoma macrotis*), 52, 54, 195
  Bryant's (*Neotoma bryanti*), 196
  Bunker's (*Neotoma bunkeri*), 196
  bushy-tailed (*Neotoma cinerea*), 46, 49, 50, 53, 54, 65, 76, 82, 84
  desert (*Neotoma lepida*), 195
  diminutive (*Nelsonia* spp.), 195
  dusky-footed (*Neotoma fuscipes*), 46, 50, 54, 70, 79, 84, 195
  eastern (in Illinois; *Neotoma floridana illinoensis*), 193, 198–201
  eastern (*Neotoma floridana*), 4–6, 8–11, 18, 26, 27, 52, 56, 75, 80, 81, 95, 97, 141
  Key Largo (*Neotoma floridana smalli*), 197, 198
  riparian (*Neotoma fuscipes riparia*), 197
  San Martin (*Neotoma martinensis*), 196
  southern plains (*Neotoma micropus*), 80, 81
  Turner Island, (*Neotoma albigula varia*), 193, 195, 196
  white-throated (*Neotoma albigula*), 81, 195
Woodrat Field Code Manual, 111, 138

MIX
Papier aus verantwortungsvollen Quellen
Paper from responsible sources
FSC® C105338
FSC
www.fsc.org

If you have any concerns about our products,
you can contact us on
ProductSafety@springernature.com

In case Publisher is established outside the EU,
the EU authorized representative is:
Springer Nature Customer Service Center GmbH
Europaplatz 3, 69115 Heidelberg, Germany

Printed by Libri Plureos GmbH
in Hamburg, Germany